THE STORY OF OUR LIVES

Homo sapiens' Secrets of Success

THE STORY OF OUR LIVES

Homo sapiens' Secrets of Success

Liat Ben David

NEW JERSEY • LONDON • SINGAPORE • BEIJING • SHANGHAI • HONG KONG • TAIPEI • CHENNAI • TOKYO

Published by

World Scientific Publishing Co. Pte. Ltd.
5 Toh Tuck Link, Singapore 596224
USA office: 27 Warren Street, Suite 401-402, Hackensack, NJ 07601
UK office: 57 Shelton Street, Covent Garden, London WC2H 9HE

British Library Cataloguing-in-Publication Data
A catalogue record for this book is available from the British Library.

THE STORY OF OUR LIVES
Homo sapiens' Secrets of Success

ISBN 978-981-124-734-7 (hardcover)
ISBN 978-981-124-852-8 (paperback)
ISBN 978-981-124-735-4 (ebook for institutions)
ISBN 978-981-124-736-1 (ebook for individuals)

For any available supplementary material, please visit
https://www.worldscientific.com/worldscibooks/10.1142/12552#t=suppl

Typeset by Stallion Press
Email: enquiries@stallionpress.com

Two drifters off to see the world,
There's such a lot of world to see![1]

To David

[1] From "**Moon River**", the Oscar-winning song from the 1961 movie *Breakfast at Tiffany's*; composed by Henry Mancini, lyrics by Johnny Mercer, originally performed by Audrey Hepburn.

Contents

Preface

Recently, I had the following conversation with one of my granddaughters, who was very amused by my age.

"Gran," she said, laughing. "You're *old*!"

My first instinct was to disinherit her, but she's five and extremely sweet, so I forgave her.

"I am."

"So you're going to die?"

"We all die eventually."

She thought about it for a moment. "And you'll be in a grave?"

"Probably."

She thought again. "What will be written on it?"

I smiled and said: "It was one hell of a ride!"

Which, thankfully, it still is.

They say the apple doesn't fall far from the tree, and my existentialism-curious 5-year-old granddaughter is a great example. Just like her, I am constantly exploring these questions: Who am I, what am I doing here, how much am I enjoying it and how long will it last? At times, it gets me into trouble, but most of the time it is great fun. It makes large chunks of my time feel like an adventure.

Ever since I can remember, I have been stimulated by diving into questions about the world and my place in it. How is everything I see connected to *me*, what is my identity, and more specifically, which parts result from the fact that I am a human being, which parts results from me being a *specific* human being, and what do I *as that specific human being* want from my life?

If these questions resonate with you as well, then you may find interest in what this book is all about. Us. Or, more accurately, who I think we are.

The answers I constantly search for are always based on questions in a multitude of areas. "There is no shame in not knowing," my parents used to preach. "Only in not making the effort to learn." And, with that, they made sure to expose me and my siblings to every type of human achievement. Surrounded by a multitude of experiences, books, music, performing arts, labs and so on, I devoured it all as an extremely tasteful meal that is exploding with thought-provoking insights, realizations and understandings. I am forever indebted to them for an amazingly rich head start.

I chose to major in biology, not stopping until I received my PhD in molecular biology at Israel's acclaimed Weizmann Institute of Science. I was constantly busy with many other areas of fascination as well, such as literature, philosophy and music.

And, education. I have always had some sort of job that involved me being an educator, whether I was a summer camp counselor or a graduate student mentoring other students. After completing my PhD in 1991, I decided to combine my love of science with my love of teaching and become a professional science educator. It was a choice that came from two strong currents in my personality:

1. I work well when I'm highly motivated,
2. I adore having an audience.

Let me explain.

One of my favorite quotes is by Charles M. Schultz: "I love mankind. It's people I can't stand!" I totally agree. Trying to bridge the gap between the two is what my motivation for being an educator is all about. Since I am convinced that science and technology are among the greatest achievements of mankind, the best logical tools we have to understand and function in the world, that's where I place my bets.

As for having an audience, it's a streak that I believe all educators worthy of their profession must have: The desire to be on stage. To some extent, we are all frustrated actors, enjoying having the attention of a captive audience. Think of it this way: In the traditional education systems, you walk into a room full of people, close the door — and they have no choice but to follow your lead for at least 45 minutes. Delightful.

While most of my peers with a PhD desired to become researchers or to work in various industries, I preferred working with students of different age groups, while learning how to communicate science and develop creative ways to learn it effectively.

Thus, science education it was. I joined the staff of the Science and Technology Education Center at Tel Aviv University. It was the last decade of the 20th century, and I spent it teaching, learning, developing curriculum, publishing books, creating educational projects, establishing learning centers, training teachers and educational leadership — you name it.

As the third millennium developed, I spent 6 years as the CEO of the Wolf Foundation, awarding the internationally acclaimed Wolf Prize to the world's leading scientists and artists. Since 2017, I have been the CEO of the Davidson[2] Institute of Science Education at the Weizmann Institute of Science. To this day, I continue to be an obsessive educator and writer. I have had the honor of doing two TEDx talks. All in all, for the past 30 years, I have been doing almost anything you can think of in science education.

During this entire time, my husband David has been my most enthusiastic partner and greatest supporter. A nature-loving amateur photographer, his insights have always found ways to be expressed by his camera, revealing the beauty and excitement that

[2] **The Davidson Institute of Science Education** is the educational arm of Israel's internationally acclaimed Weizmann Institute of Science. It develops and disseminates science programs, from scientific literacy for all to scientific excellence and development of the future scientists of Israel.

this planet is blessed with. From tracking polar bears in the Arctic Ocean in the north to chasing penguins in Antarctica in the south, from admiring the natural riches of the Galapagos archipelago through those of Africa and Asia, we have been two drifters chasing the same rainbow's end. All the photos in this book are his.

This book is a result of **my** journey. It is the **story *I* tell**, rooted in professional experience and understanding of science, technology, education, culture and the interactions between them.

So — How Does This Story of Mine Develop in This Book, in a Nutshell?

The first two chapters will acknowledge that we are an organism. Chapter 1 will discuss the *similarity* of traits between other organisms and ourselves. Chapter 2 will focus on the basis of what makes us a *unique* organism. I will discuss five major characteristics that have co-evolved in our species. These are the "**big five**": An exceptional, game-changing combination of co-emphasizing abilities, displayed as such by humans alone.

The rest of the book will embark on a journey to four[3] of the major aspects that have become pillars of homo sapiens' identity: Intelligence, technology, education and science. I will define, discuss and introduce a paradigm of how we should perceive them and their roles in our endless journey toward dominating our lives and the world, for better and for worse.

Chapter 3 will dive into one of the most important features found diversely in the living world, the one we boast that we display in extreme: Intelligence. After laying the ground to appreciate what we understand about intelligence in general, Chapter 4 will focus on one of its most important forms: ***Technological intelligence.*** I will define technology in general and technological intelligence in particular. The chapter will lay the argument that technological intelligence is unique to humans alone, and a major driving force of our success as a species.

[3] The book will focus on these four. Others will be mentioned throughout the text.

Working with the "big five", technological intelligence is part and parcel of two additional developments that are part of human success, introduced in Chapters 5 and 6: ***Education*** and ***Science***.

Like many previously discussed traits, learning and teaching as well can be found throughout the living world, but only humans educate. Chapter 5 will introduce the difference and its impact on our achievements as a species.

Finally, Chapter 6 will crown the most recent and influential human development of all: Science. More accurately, modern science, as connected to every one of the previously discussed aspects, especially and most importantly, its connections with technology and society and its place in education as a fundamental tool for designing the present as well as the future.

As I set out to write, I followed my own "ten commandments":

1. This is a popular science book. Hopefully not too simplistic, hopefully not too sophisticated. It's for people who want to read about ideas and dilemmas, think about them, and then go and explore further on their own.
2. It will not be an oversized book. Nobody needs another doorstopper.
3. I will try to avoid being repetitive and redundant. There is no need to retell what many others have already elucidated upon previously, such as evolution, human history, early hominin cultures and other excessive definitions. They are plentiful and easy to access even without me. The "Further Reading" section at the end of the book can help.
4. It's my personal version of the story, so my personal stories are part of it. Others may tell different stories, leading to different perspectives and conclusions. That's what diversity is all about.
5. Everything we perceive as unique to humans originated from other organisms. The examples are abundant and growing, as is our understanding of them. The first chapter will lay the grounds for this idea with numerous examples. There are many,

many more I could choose, but this is where "commandments" two and three come in.

6. There is no doubt that we are one of this world's organisms. Having said that, we are also very different from the rest of them. This difference is a result of the process of co-evolution and amplification of traits. I have focused on what I call "the big five": The combination that is the leading "secret" of our uniqueness.
7. Understanding the basic ideas of the systems theory is crucial for understanding, well, everything. It's the basis of understanding who we are, what intelligence is, what our technology is, why multi- and interdisciplinary understanding is so important, and how we and all the components of the world are built and work. It is a thread that underlines the entire text, and I will dwell on it in Chapter 3.
8. The world — nature, reality, the way they work — exists without us. Science and technology don't. So, mine is a very human-centered approach of the subject.
9. We are all frustrated by our education systems. Science is still struggling to be accepted. Dealing with these issues, Chapters 5 and 6 might seem somewhat more discouraging than the rest. Sorry for that.
10. Most important:
    The world is a beautiful, exciting and inspiring place.
    Handle with care.

Acknowledgements

It takes a village to write a book. Mine would not have been written without the encouragement and insight of many people whom I would like to thank.

First and foremost, I would like to express my deepest appreciation to Rochelle Kronzek from World Scientific Publishing. Shelly, without your endless support, professional guidance and unwavering trust, I wouldn't have done it! You are a fountainhead of motivation and inspiration, and I am lucky to have you as a mentor and friend.

To my colleagues and friends, Zohar Menshes and Dr. Yossi Elran, who encouraged and drove me — in their words — to "sit myself down and write".

To my editors from WSP, for their careful and attentive supervision.

To my family — my children and grandchildren, a source of inspiration, love and laughter.

Finally, most of all, to my partner, the love of my life, best friend and companion, David. You are the wind beneath my wings, and I know I'm the wind beneath yours. We have been flying together to discover and build our world for more than 40 years. There's still a lot of world out there for us, and there is no one else I would like to discover it with. Here's to the next 40!

You're water —
We're the millstone.
You're wind —
We're dust blown up into shapes.
You're spirit —
We're the opening and closing of our hands.
You're the clarity —
We're the language that tries to say it.
You're joy —
We're all the different kinds of laughing.

Mawlana Jalal-al-Din Rumi, 13th century

Chapter 1
Origins

I was standing on the bridge of the Ocean Diamond, a beautiful expedition vessel. It was the perfect position to observe how the captain and his crew navigate the ship through the South Atlantic Ocean, on our way to the Antarctic waters. I was there by a combination of mere coincidence and a modest zest for adventure: A few years earlier, at the tiny airport of the Galapagos Islands, on our way back home after two glorious weeks, my husband and I met a couple of photographers. A conversation began, and I grew envious as I listened to the global adventures they were lucky to enjoy. It seemed like they had been everywhere on the planet, while we were mostly stuck at home washing dishes.

I could have listened to their stories forever, but it was time to board our planes. A minute before we parted, I asked them one final question.

"If you had the chance to go to only one place on earth," I asked, "What would that place be?"

They didn't pause to think or blink. "Antarctica via South Georgia Islands," they both exclaimed immediately. "If it's the only place you go, that's where you want to be. There is nothing like it."

My husband and I looked at each other. Without speaking, we had just decided where our next trip would be.

At the end of 2014, we were on board a ship, led by a crew including two doctors, several scientists on their way to research penguins and climate change, a few dozen excited world travelers, and us. A biologist and her amateur photography-loving spouse, we were anxious to see with our very own eyes the last frontier of the world — Antarctica, the white continent of the South Pole (Photos 1–4, pp. 163–165).

Surrounding us were endless shades of black, blue and white. The water was black as oil. Here and there, strips of darker and lighter blue currents were noticeable, some laced with white or light gray foam. The sky that rose from the horizon felt like an extension of the water, with yet more shades of blue, gray and white. "The water beneath the sky and the water above,"[1] I recalled the perception that the ancient Hebrews had about the world. It was a breathtaking sight.

"Come here, I'd like to show you something," the captain's voice cut my thoughts. He was pointing at one of the radar screens. The bridge had several of them, indicating to whomever understood them where we were, how fast we were sailing and toward which direction. For me, wherever we were was an endless, stunning pallet of joyous blues, laced with the occasional passing of birds, whales and other sea organisms.

"Look at the screen," he said, "and tell me what you see."

The screen was empty. I waited a couple of minutes, silent, wracking my brain. What am I supposed to see? Can you see whales on this thing? One fish, two fish, red or blue fish?[2] I looked again.

Still empty.

"Well?" he insisted.

Nothing. Finally, I gave up. "There's nothing there," I said. "No other ships."

---

[1] **Genesis 1:7**, the exact translation from the original Hebrew reads as follows: "And God made the sky, and divided the waters which were under the sky from the waters which were above the sky, and it was so."

[2] **One Fish, Two Fish, Red Fish, Blue Fish** is a 1960 children's book by Dr. Seuss, one of the most successful children's books of all time, translated to numerous languages around the world.

"Exactly!" the captain cheered. "We're the only ones here! A great big, gigantic ocean, and it's only us all around!"

I continued to look at the beauty of blues beyond the large windows.

"Well? What say thee?" he was becoming poetic.

"It's breathtaking," I exclaimed.

The captain, a stout and incredibly nice man in his early fifties named Doug, paused and looked at me. His eyes were curious.

"Usually, when I do this trick to travelers, they become a bit melancholy, saying how small they feel surrounded by this force of nature, this huge body of water," he said. "But you…You're smiling."

"Of course I am," I said and spread my smile just a little bit more. "Here I am, floating on literally no more than a piece of metal — no offence meant to this gorgeous ship," I hurried to reassure him, "But in comparison to the 'huge body of water' you just recognized, it's no more than a tiny, man-made shell surrounded by endlessly deep, freezing water. I'm miles away from a strip of land that I can actually place my feet on without drowning. I'm dry, warm and satisfied after a lavish breakfast with eggs, cheese and vegetables. Holding a cup of hot cocoa in my hand, I'm on my way to exactly where I decided to go. All this is due to nothing more than the products of human thought and action. It's a situation and environment that I'm not really supposed to be in. When you show me any other organism on earth that can do this, I'll salute it. Until then, all of this is an ode to mankind. So I'm smiling!"

At that point, Captain Doug smiled back.

We are world-changers. This is not a bad or good declaration, just a factual one. A product of evolution, we are the organism that uses its biological abilities to design, create and harness the world. We build and implement tools and means that enable us to manipulate our environments, our lives and even our own bodies, fitting them to our ever-changing needs and desires.

We are the only species on earth that can do so.

What is it that enables us to be an organism, one of a wide variety of organisms of this planet, sharing some 98% of our DNA with other living creatures, while at the same time so seemingly different and unique? Compared to us, what can other organisms do as well? Are we an improved case of "anything you can do, I can do better",[3] or is there some "secret ingredient", something that no other organism has, that makes us special? Where did the idea that we are "supreme" come from? Are we, as some say, the greatest disruptors of our planet, or are we a phenomenal success of evolution? Can we be both at the same time?

Since I can remember, I have been a curious observer of the living world and even more so of our place in it. The striking, simultaneous similarities and differences between humans and other organisms have always been an intriguing source of thought and exploration for me, be it through close observation or multidisciplinary research.

I quickly realized that the answers to various questions that arise when comparing humans to other organisms are based on a wide range of understandings. While most are the result of scientific research, they are not limited to science alone. Whether we rely on genetics, ecology, paleontology, sociobiology, neuroscience, behavioral and communication studies or many disciplines altogether, our answers are highly influenced by pride and prejudice — our perceptions, cultural background and beliefs. The apparent dominance of humans over major forces of the world, as well as our growing ability to harness these forces, is mostly used for the benefit of our life quality and longevity. These are constant feeders of the unwavering, not to say stubborn, paradigm of humans as a complete and separate entity from other organisms, or at least much "better" than them.

The accumulating knowledge in the different disciplines we have created — sciences, social studies, psychology, anthropology and so on — uncovers different aspects of human nature. Biologists, for example, discuss large brains, upright posture and the ability to develop language. Some add complex tool building and unique sexual behavior. Psychologists dive into character and noticeable

---

[3] **Anything you can do I can do better** — a song from the 1946 Broadway musical "Annie get your gun", composed by Irving Berlin. The song is a duet between a male and female, attempting to outdo each other in increasingly complex tasks.

skills, such as abstract thinking, morality and self-awareness. Sociologists claim that our community-building behavior is unique, expressed in unmatched high-level organization, sophisticated means of communication, responsibility and welfare structures. Educationalists talk of knowledge transfer, motivation, individualism and socialization.

Close observation of the world we live in reveals that while each of these ideas has merit, when standing on their own, none of them is sufficient enough to be considered as *the* one, the ultimate answer as to whether and why we are unique.

Settling with "it's complex" sounds trivial, but that's exactly what it is. The complexity of the living world is no less than stunning, and the **origins** of any answer given as the reason for our so-called "preeminence" can be viewed, at least to some extent, in other organisms as well. In simple words, **the dynamic, diverse world of living organisms is filled with what can be considered as "previews" of human traits.**

It is only fitting that before diving into a discussion focused on ourselves, we roam through some of the beautiful diversity of the living world and find some of the mannerisms we seem to have in common with other organisms. Bearing in mind the enormous body of examples that can be found to make the argument, I have selected hardly a handful, concentrating on some of the features and behaviors that can be viewed as forecasts of what has developed through evolution to a much higher extent in humans. These include courtship, reproduction and family relations; housing, community-building and communication; ethics, scruples and the perception of time; centrality of food and agriculture; and culture, learning and education. If you wish to learn more or find references to the arguments and examples, please go to the "Further Reading" chapter, on page 153.

Explained through human eyes in a storytelling fashion, the goal of describing and making sense of what we see in connection to our world sometimes leads to semantics of the teleological kind — regarding nature in general or organisms in particular as knowingly behaving or choosing desirable alternatives in view of goals to be achieved. I strongly emphasize that these are used for narration purposes alone, and **by no means** should any of the observations and accompanying arguments be understood as a result of any kind of natural purpose, intelligent design or intentional power of any sort.

Without further ado, let's begin our journey.

Let's Talk About Love[4]

"Dancing cheek to cheek"[5] is romantic, and we obviously aren't the only ones who enjoy it. Courtship, mating and copulation are the bases needed for continuous survival, as well as one of the most intriguing, diverse arrays of behaviors that can be observed in mammals, birds, insects and even reptiles. Each observation offers a unique perspective, and together they create a holistic rainbow of color.

The two albatross birds were surrounded by thousands of breeding pairs, some standing on the edge of a cliff in the Falkland Islands with the ocean roaring beneath them, others mounted on bulky nests made of grass. Indifferent to their surroundings, they shut out their peers, other species and the few humans that were present. These two concentrated only on each other. Looking each other in the eye, they began to display a beautiful beak-to-beak clicking, bill clappering and sky-calling courtship dance (Photos 5–7, pp. 165–166). Prancing in place while fluttering their wings and tails, they went on and on with the entire ritual. No wonder they were so engaged in each other: Once they mate, the bond is lifelong. No one wants to make the mistake of choosing a wrong life partner. The more I was enchanted by this beautiful ritual, I was also somewhat frustrated to learn how close to us they really are:

[4] **Let's talk about love** is a 1997 Album by Canadian singer Celine Dion.

[5] **Dancing cheek to cheek** is a 1935 song written by Irving Berlin, first performed by Fred Astaire in the movie Top Hat, and later performed by other artists, such as Frank Sinatra and the duet by Louis Armstrong and Ella Fitzgerald.

After they become a pair, the courtship phase is over and the rituals are slowly and mostly abandoned. They're too busy doing laundry.

Nazca boobies in the Galapagos display a very similar dance, beginning with "sky pointing", a ritual that the male produces when he stretches his beak toward the sky. The newly formed couple dances to indicate where their nest will be, usually on rocky terrain along the ocean cliffs. After all, they paid for a room with a sea view (Photo 8, p. 167).

The living world is filled with courtship and mating rituals. Sexual selection — the selection of a spouse with whom to produce offspring — is a specific, albeit leading, case of natural selection. Through a mixture of genes and enhancement of combinations, it is the best agile model in nature that copes with the most important characteristic of environments: They change. By creating genetic variety, sexual selection is the major driving force of the development of species.

In most species, males are the ones who court the females. Dressed to the nines, serenading, offering gifts and combating each other for the attention of the ladies, the males invest resources, energy and ingenuity to attract a spouse. In some species, the display of extravagance by the males is so explicit that scientists suggest that the extraordinary morphology and risk-taking have developed as an evolutionary advantage, signaling greater biological fitness of the mating male.

The peacock train and its ornaments are thought to be related to sexual selection. The male peacock "wastes" energy on a glorious, attractive though very heavy train, showing his ability to survive and overcome predation to lure the gray, not to say plain, female. "See how strong I am?", his seemingly pointless train suggests, "Choose me and we'll have the best kids *ever*!"

Similarly, of all the exotic Galapagos seabirds around us, there was no mistaking where the frigate males were: They displayed handsome red throat pouches inflated like birthday balloons, pointing their bills to the sky where the females flew, producing vibrant drumming sounds, as if calling, "Look at me! See how splendid and brave I am, sitting here in the open! Our children will be just as successful, come and be my mate!" (Photos 9 and 10, pp. 167–168).

Bachelor herds of giraffes interact in what is called "necking", a very different interpretation of the romantic necking that humans display. What seems to be a delicate, long-neck tango between two young male individuals is actually a life-threatening combat on their way to resolve ranking, to be recognized later as the establishment of mating rights (Photos 11–14, pp. 168–170).

The popular phrase "Love can give you wings" becomes literal when every year, following the first rain of autumn, the air in Israel fills with flying ants in what looks like a huge festival of sex and rock-n-roll: The rain is the signal that young male and queen ants are waiting for. Living in their parent colony and preparing for "when the time is right" and environmental conditions are in favor for their nuptial flight, they embark on a huge celebration of intra-colony mating.

Although we like to believe that we are the only species that has sexual intercourse as a source of fun and recreation, bonobo chimpanzees display recreational sex as well. We are, however, the only species that has developed the ability to harness its own reproduction and the ability to control the number of offspring to our hearts desire — Not a trivial matter when one considers that the main goal of all organisms in the living world, from viruses to humans, is to reproduce. Besides us, every other organism is saying, "Let's create offspring. Lots of offspring. So at least some will survive. Let's duplicate by the hour. Or lay 40, 50, 100 eggs and hope for the best. Or, if we can't have lots of them, then at least let's develop ways that ensure that they are well protected in our uterus and take care of them after they are born. Let's do anything that will help our family, hence our species, to survive and thrive."

Courtship and reproduction aren't the end of the survival story. No more than five minutes after landing on the shore of South Georgia Islands, to our delight and wide-eyed surprise, a female sea lion gave birth to a cute black cub. Daddy was standing nearby, not too close, but obviously alert, watching over Mom and their newborn child. She looked exhausted as she caressed her cub, who was still wet from labor fluids and lay near the empty but fleshy, somewhat bloody placenta. The cub raised its head toward her and started to wiggle clumsily. It was enchanting (Photos 15 and 16, pp. 170–171).

The cuteness of the cub was no surprise: One of our first reactions when we see a mammal baby involves cooing, the desire to hug it and take it home with us, forgetting that this cuddly, soft-furred teddy will become a huge and aggressive grizzly bear, or that this cute little kitten-like round face with wide, begging eyes and adorable expression will soon grow up to be a fierce cheetah, who will easily forget that we once fed it and are by no means meant to become its food. Evidence shows that mammals recognize and are influenced by newborn baby features even between species, at which point parental protective instincts may kick in and cause a female leopard to care for a baby baboon instead of devouring it altogether. Based on these instances, some argue that this "baby face syndrome" has developed in mammal cubs as a result of evolutionary advantage: Mammal newborns cannot survive on their own, requiring a lot of nurturing care before they are independent of their parents. Mammal parents invest time and energy in feeding, grooming and protecting their young, while many times endangering themselves during the process. Not the best parenting deal in town. Mammals can be found to neglect offspring that are less strong, or have less obvious "cuteness" features. If you're not cute, forget it. You're too much trouble.

But, this cub was definitely cute. I stood there, enjoying the calm scene of the new sea lion family.

Suddenly, dropping wildly from the sky, two large, aggressive and loud brown skua birds landed next to the proud mother and cub, flapping their wings and aiming at the small pup.

Mom became frantic. Her exhaustion disappeared. Covering her infant and calling out to Dad. He immediately came to the rescue, roaring ferociously, using all his energy to battle the birds who were endangering his family. After a long, violent encounter that included loud voices, snapping of teeth and bills, biting and clashing of wings with fore flippers, parental love won.

Acknowledging that Mom and Dad mean business and will attack to death, the skua managed to capture the placenta (Photos 17–19, pp. 171–172). With the bloody piece of flesh dangling from their beaks, they were driven away.

When they were high in the sky and far from sight, peace returned to the family. I could have sworn that the look in the cub's eyes was made of gratitude and pride: "*Did you see that*?? Yup, these are *MY* parents!"

Our House is a Very, Very, Very Fine House[6]

African weavers are exactly that: Weavers of beautiful nests, cylindrical in shape, with round, tubular entrances. Like many other species, male weavers are usually beautifully colored in yellow, red or orange.

But, courting a female using beauty isn't enough. The lady wants to make sure that her suitor can provide a fitting home, where their offspring will be safe and merry. So, he weaves nests made of grass, twigs and other plant material that can be used to create a basket. Using only his feet and beak, the nests are a skillful work of intelligent, natural art (Photos 20–22, pp. 173–174).

Building the nest is only part of the job. In fact, the male is expected to be a professional constructer, building several nests before the female decides which of them is to her liking. Only after she is satisfied with the nest of her choice will they mate and breed, leaving the unchosen nests abandoned and empty.

Although it seems like a huge waste of energy, the evolutionary advantage is quite simple: Snakes, hawks and other predators, seeking the delicacy of eggs and chicks, are more likely to raid empty nests than actually win the prize for their effort.

So, he builds several nests. All she has to do is twerp, *I do.*

---

[6]From the song **Our House** (1970) written by British singer-songwriter Graham Nash and recorded by the folk rock group Crosby, Stills, Nash & Young.

"I want an old fashioned house with an old fashioned fence and an old fashioned millionaire", says the song.[7]

Turns out, humans aren't the only ones singing it.

The weaver family includes numerous species, one of which is the sociable weaver — highly social and cooperative songbirds. These birds don't settle for a nest per couple. Instead, they join forces and build a shared, year-round commune settled in one gigantic nest, built by members of the colony and sometimes passing from one generation of weavers to the next (Photo 23, p. 174).

Nestled on the trees of the Kalahari, at first glance, the communal nest looked like a pile of hay that someone tossed on a tree. It was so large and heavy that I was certain the tree itself would collapse at any moment. A closer and longer look revealed birds chattering their way in and out, busy with bringing twigs and food. Each communal nest can be home to more than a hundred couples and their chicks, all sharing resources to meet individual as well as community needs. The nest includes chambers for each pair or family; different areas in the commune will provide different temperatures and are used at different times during the day. Members of the sociable weaver colony will have different roles, displaying hierarchies that are correlated to social status. A high social status benefits from enhanced access to resources, better breeding positions, and is shared with relatives through nepotism and cooperativeness. It is a beautiful display of sophisticated building, social interactions and community all at once, and when you think about ant colonies or beehives, it is clear that it is but one example of many.

An additional angle can take these sophisticated, hierarchical communities one step further. A common claim is that humans are unique because we build complex tools, which indeed we do. Although extensive research has shown that other species use and

---

[7] **Just An Old Fashioned Girl** was a popular song written by Marvin A. Fisher, best known in its 1956 recording by Eartha Kitt. The humorous lyrics portray the voice of a woman, setting the priorities by which she will choose her spouse.

manipulate objects to meet a variety of needs, these are very simple tools used in very simple ways, unchanged through millions of years.

However, building a communal nest or beehive requires a higher level of sophistication, as does building a dam that reengineers the landscape. Beavers are one of the few animals that modify the environment by building dams that form slow-moving ponds. These ponds serve as newly formed habitats for a wide range of animals, aquatic and on land alike, supporting a newly formed diverse ecological community. Within these ponds, beavers build their lodges with one or more underwater entrances, carefully designing their living quarters that are located above the water line. Mostly built away from shore, these lodges form islands that can be entered only underwater, providing safe lodgings for the entire beaver family, implying that beavers have more than just strong teeth.

Are You Talking to *Me*?[8]

The social structure of interactions between individuals or groups, otherwise known as social networking, is old news. In fact, it is far more ancient than the age of twitter, Facebook, and even humanity itself. Identifying social networking and communication in mammals such as orcas, for example, isn't surprising. These highly intelligent members of the dolphin family not only make a wide variety of communicative sounds but each group, known as "pod", has distinctive "songs" that its members will recognize even at a distance. Communication and language are widespread throughout the living world, and are actually characteristics of life.

I was staring at the illusive, flickering lights coming and going between the trees just above the high grass. It felt like I was surrounded by Tinkerbell[9] and all her relatives. The fireflies of the Maroantsetra Nature Reserve in Madagascar were delightful to watch. These soft-winged beetles produce light within their bodies,

[8] Quoting Travis Bickle's character, played by Robert de Niro, from Martin Scorsese's 1976 movie **Taxi Driver**.

[9] "**Tinkerbell**" is a tiny, glowing fairy, in J. M. Barrie's 1904 play "**Peter Pan**".

a relatively rare ability called bioluminescence. Mostly used for mating, each firefly species has its own language — its own unique signaling system. It is used in a specific habitat, at a specific time, supplying species as well as sex identification. The signal is sent by the flying male to the females, who are usually situated on the ground or in the vegetation, waiting for the show. When the female notices that the male show has begun, she responds by sending the species-appropriate flash of her own. Then, the two mutually signal to each other as the male flies down to her. If everything goes right, a new couple is created. I was waiting with patience. It was worth it: The to-and-fro visual Morse code communication created a beautiful festival of lights in the grass.

While watching a group of playful dolphins in the waters of the Galapagos, there was no mistaking the sounds they made to each other as they circled our zodiac (Photo 24, p. 175). They were clearly a community, a result of mutual communication and work that begins before birth. Dolphin females teach their unborn babies a "signature whistle" — sounds made by individual dolphins, by which they identify one another. Right before birth and up to 2 weeks after, female dolphins will increase the frequency of their whistling so their calves can learn to identify and find them.

Chattering, social roles and organization, coordination and even espionage are found in places that we would least expect them, such as between different types of trees, fungi and other organisms in forests. Trees have been shown to communicate and form complex systems that function as communities. They are connected underground by fungal threads. Nutrients, hormones and even alarm signals pass through these threads from tree to tree, regardless of species. Older, larger trees take care of younger trees, and when one tree is attacked, chemical warnings are sent between them.

"Communicate with thy neighbor — chemically" — is a common trait in the microbial world as well. It isn't the best idea, though, since anyone in the environment can "read" your message, eavesdrop and counter-react, exposing the communicators to potential harm. So much so that some bacterial species developed their own evolutionary version of the code, packaging the chemical messages within cell membranes like a message in a bottle. Some microorganisms organize as "herds", creating flexible links between individuals, allowing them to move synchronously in large clusters, communicating and "hunting" together for nutrients to feed on. When nutrients become scarce, cells coordinate their movement and predation strategy for the benefit of the entire community.

And, communities they are: A wide variety of microorganisms, such as certain types of Amoeba, Myxobacteria, *Pseudomonas fluorescens, Bacillus subtilis* and more, form extremely large groups that behave collectively in ways that impact the individual members. These microscopic communities are characterized by structural and genetic heterogeneity, complex communication and social interactions, including "safeguards" from toxins, UV light, dehydration and other organisms.

Research increasingly suggests that molecular "conversations" take place among members of a broad spectrum of microbial communities, and also between a variety of microbes and host organisms. There is the emerging sense that microbes interact in complex, diverse and subtle ways that we have yet to fully decipher. One thing is clear: We definitely didn't invent manipulation of communication, otherwise known as gossip.

If You Don't Like My Principles, I Have Others[10]

We like to believe that humans are the species of morality. We have an explicit sense of awareness, a clear understanding of the profound meaning of "I am", differentiating "me" from "you". We claim to know the difference between right and wrong. We act according to cultural values that we define as meaningful, designed for what we consider as the benefit of our well-being. We openly declare that we are the only living creature that has a code of ethics.

---

[10] A saying ascribed mainly to 20th-century American comedian and movie star, **Groucho Marx**. He was not the first to say it — politicians in various areas of the world have used it as early as 1873 — it fits well with a quick, witty sense of humor, and has been attributed to him since the 1980s.

Or are we?

I was standing on a snowy hill in Antarctica watching chin-strapped penguins build their nests. They do so by collecting stones and small rocks from the ocean shore, which they carefully choose and carry uphill in their beaks, meticulously placing and lining each stone in circles to create the structures where they breed. It was a freezing day, the wind was blowing hard and I watched how they slaved, struggling to make their way through the deep snow and all the way uphill to the nesting area. Each placed his stone in his own specific nest, then went all the way back down for yet another stone. It was a harsh, laboring endeavor, and as I found difficulty in every step I made against the freezing wind, I could do nothing but admire their commitment and perseverance.

As I continued to observe them, I noticed that one of the penguins wasn't following the rest. He just stood there, watching the working congregation. To my surprise, each time they weren't looking, he ran up to some other penguin's nest and, showing no hesitation or remorse, quickly stole one of the stones and hurried off with it to his own nest. A couple of times, when other penguins were around, he began walking downhill with them, making sure he was falling behind either by being extremely slow or by suddenly flapping his wings as if something was bothering him. When the gap between him and his peers was large enough, he turned back and returned to his mischief. Whenever he thought one of his peers might be looking, he suddenly stood again, looking around while avoiding any eye contact, as if saying, "I'm just standing here minding my own business". His behavior left no room for doubt: He knew exactly what he was doing, and he knew that it was wrong.

The little thief continued to manipulate his way around like this for quite a while, until finally his peers caught him red-handed. A full-fledged racket burst through the entire flock. Only a penguin will understand what they were squawking at each other, but it was obvious that they were furious. The crime was followed by harsh punishment: All the stolen rocks were placed back in the nests they came from, and the thief was forced into solitude, ignored or sneered at by his peers (Photos 25 and 26, pp. 175–176).

Alas, manipulation for one's benefit and the scruples that rise from it — signs of moral standards — are not unique to homo sapiens alone.

The Invention of Time

The youngster walked up to his passed away mentor. Bowing his head, he raised his trunk and placed it on the corpse in front of him with a gesture surprisingly gentle, especially considering it came from a very large elephant with very sad eyes.

His obvious grief nearly broke my heart.

A group of his fellow elephants stood around him, waiting patiently in respect. He stood there for a few minutes, then slowly walked backward into the line.

It was one of the most painful, sincere displays of mourning I have ever witnessed.

For more than 24 hours, surrounded by the harsh landscape of the Savuti Nature Reserve in Botswana, I was fascinated by the long parade of elephants as I watched them walk up, stand vigil and pay their respects to the fresh corpse of their peer that was hunted and killed by a pack of lions. The lions lay nearby, their bulging bellies throbbing as they digested their meal. The lions will hunt again only after they have devoured their prey to the bone, a process that would take approximately three more days. The elephants were calm to be so close to the hunters, knowing they are currently safe. Even when one of the lions walked over to the corpse for another helping, the elephants remained indifferent, focusing only on their lost friend. It was an evident display of recognition that there is time to grieve before the next hunting, a recognition that requires reason and understanding of the concept of time (Photos 27 and 28, pp. 176–177).

Obviously, these African elephants had never heard of the argument claiming that humans are the only organism that grasps the existence and difference between past, presence and future, or in other words, time-oriented cognition. Evidence is accumulating suggesting that other organisms, such as birds and mammals, show predispositions of this capability as well. Moreover, there is initial evidence that other organisms, such as orangutans, can exhibit displaced reference — the ability to communicate something that occurred in the past or will occur in the future, generally believed to be unique to humans alone.

Time-oriented cognition is indeed advanced in humans. We readily and continually draw conclusions and behaviors from the past and make plans for the future. These abilities are based on abstract, long-term thinking skills, on explicit and implicit intentionality and on a well-developed sense of time. Thanks to these skills, we are able to actively take measures that affect our future. Like other characteristics, these, too, have developed and been refined throughout evolution.

Food, Glorious Food[11]

One of the very first and most basic needs is food. Every social gathering, from the celebration of birth to the mourning of death, is accompanied by food. "We fought, we won, we eat; we fought, we lost, we eat" is a common Jewish joke, describing our multitude of feasts for every occasion. We use food as symbols, as a source of comfort and as gifts, sometimes so precious that we will build a temperature-, pressure- and humidity-controlled environment to take care of an exclusive type of melon, pear or grape, as anyone who visits a high-end supermarket in Japan can witness.

We aren't the only species to go out of our way when treating food. It is something we learn from nature. The beautiful bee-eater doesn't settle with being dressed to the nines to lure a female. He makes sure to bring her the gift of food, catching insects while soaring through the air and then bringing his prey to the female of his liking. If the food he offers is to her liking, she will mate with him (Photo 29, p. 177).

---

[11] **Food, Glorious Food** is the opening song from the 1960s musical **Oliver!**, based on Charles Dickens's 1838 novel **Oliver Twist**. Music and lyrics written by Lionel Bart, the song is sung when the orphan boys are fantasizing about proper food while receiving their daily gruel.

We hold food and beverage near and dear to our hearts, continuously developing new sources and ways to treat and care for our foods, expanding the variety of offerings, colors and tastes for our palate's delight and our buffet's richness. The most common argument regarding our deep relationship with food claims that we are the only organism that actively grows and manipulates its own food. We began mastering this ability some 12,000 years ago, when we started to domesticate animals, plants and developed farming and means to supply water. We developed agriculture, an entire discipline of methodologies and knowledge to supply food, lots of food and different types of food for ourselves. Search for edible plants was replaced by growing them. Search for water sources led to creating reservoirs and water systems, and the development of a multitude of beverages is a story within itself.

Hunting was diminished by domestication of animals as a source of meat. It freed us from the need to hunt and collect whatever we could find, reducing our dependence on forces of the environment and harnessing them in our favor instead. Our agricultural ability to grow our own food has been a major driving force in the development and growth of human settlements, commerce, communication and governing, or in short of civilization as we know it.

The argument that agriculture is exclusive to humans may have been interesting if not for the well-known species of leafcutter ants. These ants carry leaves to their nests, where they will cut, chew and spit out the leaves to create the substrate on which a specific fungus will grow. The fungus is then used as food for the ants' larvae, "baby ants", while the adults feed on leaf sap.

These ants are growing food for their young.

The most intriguing fact is that the leafcutter ants don't settle with merely growing their crops, but they actually *cultivate* them — they weed, manure and prune their crops. Even if we are still to understand whether this is conscious behavior, it is undeniable that for some 50 million years, these species have been practicing what we humans have prided ourselves on as being the only creature to do for a mere 10,000 years — a simple yet planned form of agriculture.

You and Me, But Mostly *We*[12]

It was an amusing sight. In the midst of an endless bed of king penguins and their chicks, this one chick caught everyone's attention when it started flapping its wings, screaming and complaining to its mother. In a set of motions and faint squawks, she desperately but unsuccessfully tried to calm him, then tried to move away from him, ignoring his continuous shrieks as he followed her (Photo 30, p. 178). Eventually, like every child with a tantrum, he became exhausted and stopped. Mom looked relieved. Three other penguins, who throughout the entire episode stood and watched from a close but safe distance, approached her after the rebellious chick calmed down. It looked like they were consoling her, as if to say, "Kids will be kids. Don't worry, he'll soon grow up and move out of the house."

As much as I enjoy talking about myself, I was frustrated when my search for "Animal personality" yielded references that were all about, well, me: "Which animal are you?" or "What animal shares your personality?" were the leading titles I found.

For a moment I was distracted from my original goal, happily cooperating and thinking to myself — ok, am I a deer or a whale? Maybe a dolphin? An arctic tern will actually be the best choice for me, flying from one pole to the other, beautiful and elegant with my snow white feathers, red beak and black head. I realized, again, how easy it is to become self-centered and drift down that lane. I wondered if any other animal, when presented with the same question, would chose to be a human being.

In spite of what we believe, it's not all about us. Animals' personalities should have been a straightforward, obvious notion to recognize, regardless of humans. Anyone who has raised a pet, taken care of an animal or observed the behavior of a group of animals knows that, just like humans, different animals have different personalities and behaviors.

---

[12] Paraphrase of the song from the 2011 musical **Book of Mormon** — "You and me, but mostly me", a parody on individuality vs. friendship.

I was watching the two youngsters who were rolling on the ground, obviously teasing and striking each other with playful and gentle strokes. One offered a branch to the other, but withdrew it a second before his friend reached it, jumping aside and waiting to see the reaction he provoked. As they continued, a third friend joined in and it wasn't long before all three were engaged in what resembled a game of catch. They reminded me of my own children, and if they hadn't been orange Diademed Sifaka lemurs, I could have sworn they were actually laughing (Photos 31–33, pp. 178–179).

Other lemur species throughout the island of Madagascar displayed similar behaviors, from teasing to snatching objects, such as my hat, notebook and water bottle which they took, explored and then gave back to my utter surprise (Photos 34 and 35, p. 180).

Preliminary evidence suggests that, like humans, other primates exhibit playfulness, enjoy games, laugh and even tease. Great apes have been seen to display playful behavior described in human infants, such as mocking and disrupting others' activities. This kind of action involves differentiating between "me" and "you". It includes a complex socio-cognitive understanding of expectations, then expressing this understanding through personality-based behavior.

Respect and recognition of others sounds like a lesson we all need to take at school, and there was no reason for me to think that lemurs attended such lessons. When the large Indri lemur looked me in the eye, held my hand in his paw and aimed to take a banana slice between his quiet intimidating teeth, I was somewhat nervous. In an extremely careful and gentle manner, his made the softest of gestures I had ever experienced (Photo 36, p. 181).

Friends are something we all need and yearn for. Three king penguins standing on a hill sounds like the beginning of a joke, but in this case, it seemed more like a small, heartbreaking triangle. While two were looking at each other, using their beaks and making some sounds that created communication, a third walked up, looking for companionship. While he tried to join the conversation, they just ignored him. In what seemed like a display of frustration, he moved closer, trying to draw the couple's attention to sounds and movements he made. Unconvinced, they continued to ignore him, until finally one of them made a loud sound toward him and abruptly turned his back on the desperate, friend-seeking penguin, taking his comrade with him. The insulted penguin sounded something, watched their backs as they walked away, then turned around and left in a heartbreaking, head-bowing sad gesture. The only thing missing was the song "lonely, I'm Mr. lonely"[13] in the background (Photos 37–39, pp. 181–182).

We've established that fun and games were not invented by us, but it was especially entertaining to see Gentoo penguins run wobbling downhill, falling on their bellies and "skiing" all the way down, then getting back on their feet, wobbling back uphill only to ski back down again. These penguins weren't "skiing" on their bellies — an action called "tobogganing" — to merely move from place to place. They were obviously having fun. I could have sworn they were looking at us humans, teasing, "For us, ski season is never over!" (Photos 40 and 41, p. 183).

Similarly, I was utterly amused to see a group of Ugandan Baboons use a solar panel as a slide, going up and down like children in a playground (Photo 42, p. 184).

Animals develop cultural behaviors as well. In a nutshell, culture is the transfer of specific traditions through social learning in a specific group, between its peers and generations alike. Culture is the manifestation of diversity, a result of the way different groups of a species display the same kind of activity in a variety of ways.

Until recently, culture was considered as an exclusively human phenomenon. The dichotomy was clear: Animal behavior is a result of instincts and imprinted processes, while anything considered as culture — such as art, music, fashion, science and so on — is a marvel of humans alone.

In the past decades, researchers of animal behavior have been discovering more and more evidence of the cultural abilities of other organisms. From chimpanzees creating ornaments to song dialects in birds, other organisms demonstrate behavioral systems that are transferred through social learning within the group. Like

[13]From the 1962 song by American singer Bobby Vinton, "**Mr. lonely**".

other areas, the development of social interactions leading to culture is an evolutionary continuum in the living world.

We are more like other animals than we care to acknowledge. All you have to do is look for it.

My Family, Among Other Animals[14]

An endless network of relatives, life on this planet is a kingdom that has it all. There is a swelling body of evidence proving that animal emotion, individuality, personality and behavior are aspects to consider, primal to everything we know about ourselves. They are as important as genetics and physiology are for understanding a world full of organisms that are constantly competing, cooperating, adapting and ultimately evolving in a continuously changing ecosphere. Sociobiology, ecology, animal behavior research and so on have revealed character diversity leading to behavioral and simple transmission in a variety of organisms.

Examples are abundant and accumulating: Interactions in a group of baboons dominated by a baboon of violent character changed dramatically when a new baboon of gentle character took his place. Capuchin monkeys teach their young which food to prefer based on parental experience. Adult female Orcas mentor adolescent females, teaching them infant-caring, training them to be future mothers. Humpback whales have been reported to spread amongst their species a new feeding technique within a few years. Chimpanzees learn from each other how to play games and modify objects as tools, passing patterns from one generation to the next through observation, imitation and practice. Dolphins have been seen to teach each other shelling, a hunting technique utilizing large conch shells to catch fish, a skill they appear to observe and learn from one another.

Learning from the experience of others increases the ability of the entire group to be flexible and adapt to changes. We learn not

---

[14] **My family and other animals** (1956) is the first book of the 'Corfu trilogy' written by British naturalist Gerald Durell, in which he weaves the story of his family in the stories of animals on the island, developing the understanding that we are all part of one, large biosphere.

only from other humans but also from other species, such as the case of induced labor by females in Africa. When birth comes near, pregnant female elephants are known to eat a tree of the Boraginaceae family, "forget-me-not's", used also by female women to induce labor. Documented cases of plants being used by animals in what seems to be medicinal purposes are extensive: Birds, bees, lizards, elephants, and chimpanzees all eat things that make them feel better, prevent disease, kill parasites or aid in digestion. The traditional Dragon Boat festival in China marks what is considered to be a bad time of the year, when evil spirits, poisonous creatures and illness might invade and do harm. One of the customs that local people practice during the festival is hanging wormwood from their doors, believing that it will protect them against illness and health problems. Around the same time as the festival, Russel sparrows incorporate wormwood into their nests. The anti-parasite compounds in the wormwood have been found to serve as medicinal control: Nests containing wormwood leaves reduce infestation by parasites and increase the well-being of the chicks.

Whether passing knowledge from older to younger generations is based on animal instinct or on higher-order thinking skills is yet to be discovered. Having said that, there is widespread agreement that social behaviors, sophisticated cooperation, communication, display of emotions and more, transferred from one generation to the next, are an evident basis for the development of education and culture.

The inclination to organize in cultures, design our lives and educate our children to continue them is a feature first and foremost related to humans. There is no doubt that cultural and educational behaviors have been developed to extreme complexity in homo sapiens. Our cultures are complex social structures that allow us to function as a group based on shared values, goals, norms and ideas. Human cultures differ from each other in language, religion, cuisine, social habits, music, arts, fashion and governance. The human ability to develop and transmit diverse cultural systems has played a role of evolutionary advantage, largely due to our ability to unceasingly learn and develop from one another as well as from one generation to the next, as will be discussed in the following chapters.

Albeit highly sophisticated in humans, there is accumulating evidence that culture and education originated in the living world long before hominins[15] gathered together to tell stories, create spears and teach their children how to use fire. Communication, reason and a wide variety of emotions can easily be detected when observing a variety of organisms living on our planet. Various versions of primal storytelling, gossip, love, hate and deceit can be witnessed in birds, mammals, insects and others, as they maneuver their peers to or from food, danger or mating opportunities.

It's all out there, part of the larger family of living organisms.

In a long process of evolution, all these traits continued to develop and change in primates to hominids. Some 300,000 years ago, a species developed in which this beautifully rich and diverse display of characters came together in a stunning, unique demonstration of synergy and amplification.

The species that called itself ***homo sapiens sapiens*** — Latin for "wise, wise man".

Us.

The organism that not only walks, talks and tells stories, but makes them come true.

---

[15] *Hominini* — members of the human lineage, of which today survives only one species — homo sapiens. Fossil evidence indicates that we were preceded for millions of years by other hominins. Characteristics include erect posture, bipedal locomotion, large brains, specialized tool use and communication through language.

Chapter 2
The Big Five*

In 2001, the turn of the 21st century, I was on my way with my family from Israel to the USA. We were expected to be away for at least 3 years, so it obviously seemed like a good idea that before leaving, we would visit my then 90-year-old aunt Rina. As we entered her home, our youngest son — a first grader who was very excited by the upcoming adventure — ran up to her, exclaiming, "Aunt Rina, did you hear? We're going to America!"

Rina's face lit up, she embraced him and looked nostalgically at a distant spot in her past, reminiscing, "Oh, how nice! I remember the first time I went to America. I was barely a teenager of fifteen, on my way to discover the world." Her gaze drifted down memory lane. "What an adventure it was…it took me three months to get there."

***The Big Five** — spotting all five mega-animals is a well-known challenge when on safari in Africa, referring to the five most difficult — some will say, also the most aggressive — animals to see. They are the lion, leopard (or, in some countries, cheetah), rhino, elephant and buffalo. "Only after you've spotted all five can you truly say you have been to Africa!" is the claim of local guides. Sadly, due to excessive poaching, rhinos have become a highly endangered species. As a result, in some countries, the term has been reduced to the "big four". Efforts are being made to create ways to aid recovery of the rhino population in the continent.

Panicking, my 6-year-old son turned to me and said, "Mom, *please* tell me we're not using the same airline!"

He was right. For all he knew, the distance between Israel and the USA is such that you step into a large metal box in Tel Aviv at night, go to sleep and wake up in the morning at the footsteps of the Statue of Liberty. It definitely isn't a distance that requires 3 months of travel. By the time he will have a 6 year old of his own, the distance will probably be even shorter. Not in miles, but in time and perception. It's what we humans do: we create something, then change it, then change it again.

We go about making these ongoing changes at an impressive pace. Only a few decades ago, when I was a teenager travelling to Europe, cellphones and the Internet were yet to be invented. I communicated with my parents in Israel via handwritten letters sent by regular mail, knowing that the letters I wrote would take about a week to reach them and would therefore bring them news that is already outdated. This not-so-long-ago method of communication is currently rightfully known as "snail mail". Today, I video-chat with my family from everywhere in the world, updating each other on a real-time basis. Considering that we live on an extremely biodiverse planet, with millions of different species, the fact that we are the one and only organism to do so is no less than overwhelming.

Consequently, it is regarded as evolution's greatest question of all: Where did we come from, and how did we become who we are? To be honest, it isn't evolution's question at all. Evolution knows exactly where we came from. It's *our* question, and all of man's sciences dedicate research to find possible answers. Paleoanthropologists, geneticists, molecular and computational biologists, archeologists, ecologists and other scientists are continuously researching archeological sites, fossils, DNA traits, paleoclimate dynamics, computational models and other scientific data. All this, in an effort to unravel the information hidden within, regarding the puzzle of human evolution yet to be complete.

There is widespread agreement that our story begins in one continent, recognizing Africa as the cradle of mankind. Evidence is accumulating that the origin of humans is not a simple, single-linear or even tree-like story of a one-spot-beginning, as suggested by several scientific models. Ours is probably a more complex story of continent-spread diversity, exchange and co-evolution of genes and skills between different polycentric origins of humans, known as hominins. These included different species from which our specific species, homo sapiens, developed altogether. That is, our cradle was actually a *nursery* of cradles, pulled apart or drawn together according to climate, landscape and ecosystem changes. These included widespread and regular interactions and interbreeding between our hominin ancestors across the continent of Africa, spreading to other continents and continuing their development. When addressing human evolution, network-like models are considered, including interspecies relations, gene flow, merging of branches, ongoing integration, combinations of physical features and exchange of pre-cultures. New data are constantly being discovered and analyzed, shedding light on the variety of human species, their origins, habits, connections and movement.[1]

In addition, developmental research of other organisms provides growing data supporting processes of *co-evolution*: the reciprocal evolutionary change between traits and genes within a species. Co-evolution of traits is considered a major driving force of development and complexity. It is part of the very capacity to evolve, suggesting that ours may also be a story of co-evolution of traits within our species.

It remains to be scientifically determined which model, or models, best explains the development of humans, ourselves included. Be the origin of humans what it may, close observation of five human characteristics and the interactions between them suggests that we are the result of a complex combination and enhancement of co-evolutionary style. Intertwined and inseparable, we enjoy an unprecedented amplification of abilities that work and continuously evolve together as a sophisticated system,[2] creating the so-called

[1] Details can be found in the references mentioned in the "Further Reading" section, page 153.

[2] The definition and characteristics of a "System", as explained by the General Systems Theory, will be discussed in Chapter 3.

"preeminence" of our species. These five systemic parts are **story-making, creation, communication, change** and **inter-generational transfer**. While "seeds" of each of them can be found in other organisms, their accelerated synergy in humans creates a result unparalleled in any other living creature. They are "The Big Five of Homo Sapiens". Each one of them is beautiful, and together they are no less than mind-boggling.

Let's explore them.

Story-makers

A crowd of people was gathered around the old man. He was sitting on the dust, holding an open book. The left page had a large drawing of a circus monkey, dressed in a red jacket, with a red hat on its head, and holding a drum. The right page had some text lines, and yet another drawing of what looked like a large green dragon. The man was speaking, his voice changing at moments, sometimes softer, sometimes alarmed, sometimes it felt like he was angry, sometimes he exclaimed or paused with a smile. Every now and again, he looked at the book; most of the time, his eyes were fixed on the eyes of his audience, turning his gaze from one to another. He went on and on like this for almost an hour, without turning a page. The people around him were attentive, following his every move and word, mesmerized. They reacted to what he was saying, sometimes gasping, sometimes laughing, continuously enchanted. It wasn't necessary for me to understand his language in order to understand exactly what was going on: For centuries, the *Jemaa el-Fnaa* main square and market place in Marrakesh's medina quarter has been the place you go to in Morocco when you want to hear a good story. Active mainly after sunset, the square is packed with fire-swallowers, snake-charmers, jugglers, healers and food carts. People fill the square, moving from one attraction to the next, stopping to enjoy a cobra's dance or purchase some medicinal herbs from a cart. But, the claim to fame of the square are the storytellers, the main charm sought after every single evening by crowds and crowds of eager listeners and captivated audiences. When I returned to the square the following night, the old storyteller was there again, with a different book, a different crowd, and the same enthusiastic storytelling magic.

Our ability to tell stories, create real and imagined worlds and relate to them can't be overstated. When young children believe in the tooth fairy, Santa Claus or a flying reindeer named Rudolf, it's an important stage that is necessary for the development of confidence, imagination and innovative thinking for their future. Most of us will remember the details of a good story because of the feelings it stirs in us. We laugh, cry, love and hate through our stories. Even after watching E.T.[3] dozens of times, when Elliot and his friends are trying to save him, I will still jump on the edge of my chair and mumble anxiously, "Fly! Fly away!"', and each time they finally fly, a sense of relief floods all over me again. As adults, we will still sometimes quietly wish for Mary Poppins[4] to fly down from the clouds and fix everything for us. We are willing to go to great lengths for a good story, as can attest anyone who waited at midnight on the stairs of the bookstore entrance as I did, year after year, to become one of the first proud owners of the new *Harry Potter*[5] book. Storytelling is one of the central tools we use when we invent something new, justifying our creations with a narrative we attach to them.

Storytelling is also one of the best learning tools. "We are going to study the October revolution in Russia, and its influence on the entire world," our history teacher declared at the beginning of my 9th-grade school year. "Open your textbooks and read from page 82

[3] "***E.T. the Extra-Terrestrial***" is a 1982 American film about the boy Elliot and his extra-terrestrial friend, E.T., who is stranded on earth and hunted by harming adults. Elliot and his friends help E.T. escape and "go back home".

[4] ***Mary Poppins*** is a 1964 Disney film based on a series of books with the same name, written by Australian-British author P.L. Travers.

[5] The ***Harry Potter*** book series was published in intervals of at least a year between the books, creating great expectation and anticipation amongst fans of the series, who ordered their copies a long time before the publishing date and waited in crowds in front of the book stores when the first new copies arrived. With enormous global success, it is considered as a book series that brought reading back into the lives of many children.

to page 208." After an entire semester of reading more and more texts, memorizing names, dates and events, and answering questions, our study ended with an exam. I scored the full 100 points. Three days later, I hardly remembered any of the details that I had memorized. They simply didn't mean anything to me.

The following year, I watched the movie "Dr. Zhivago".[6] It led to a crush on Omar Sharif, identification with Lara's suffering, endless humming of "Somewhere, my love" (Lara's theme) and frantic reading of more stories based on the October revolution, the events that led to it and its results, including finally listening to my grandmother's stories about her own early life in Russia.

Going back to my history textbook, I suddenly understood why this incredible, difficult, emotion-stirring and meaningful time in history seemed so dull to me only a year before: The textbook didn't tell *the story*. It gave cold, faceless and emotionless facts and figures. Thanks to the stories I read and heard, they became alive.

More than four decades later, when I arrived in Moscow for the first time, I still remembered the details of the revolution, understood what I saw in view of what I had learned, and could have an interesting, knowledge-based discussion with some Russian peers. The story made my experience richer and more meaningful.

Our drama continues when we understand that we are much more than story-*tellers*. We are story-*makers*, a species that knows not only how to tell stories but also takes the action of doing, making and building solutions to make the story we tell come true. We are creators of visions, definers of needs and pursuers of doing. We see a rainbow and dream of going to Oz,[7] envisioning what we will find when we get there long before leaving Kansas, restlessly and

---

[6] **"Dr. Zhivago"** is a 1965 film set in the Russia of World War I and the Russian Civil War of 1918–1922. The film stars Omar Sharif as Dr. Zhivago, Julie Christie as his beloved Lara, and is based on Boris Pasternak's 1957 novel.

[7] **"Somewhere Over the Rainbow"** is a song from the 1939 film "The Wizard of Oz", based on F.L. Baum's book of the same name from 1900. It is sung by the leading character Dorothy, who lives in a dull place in Kansas and dreams of a better place beyond the rainbow. When a cyclone spins her house into air, she lands in a faraway land called Oz.

relentlessly searching for ways to build the means that will enable us to actually take the journey or accomplish the vision we crave.

What are our stories and why do we need them so much?

Stories are representations of real and/or imaginary situations or events. They include participants (people, animals, creatures, etc.), perceptions and values told for a variety of reasons, such as explanation, entertainment, description and attribution, in a variety of ways. A story is a sequence of facts, events and ideas that include logic, emotions, actions and so on, creating a defined structure with coherent meaning, told by a certain party and understood by another. It can be based on truth or imagination, fiction or non-fiction, documentary — or a combination of several. We can tell a story about the past, present and future, About anyone, anything and anywhere, existing or not. Our stories are at the basis of personal and cultural development, and can defy time, space and even reality.

Part of everything we do, we adore our stories. We tell them at different levels of detail, shaping and designing their plots through a vast variety of forms — from spoken words to literature, music, art, theatre, dance, body language and so on. Like addicts, we crave narratives that function as lighthouses of value and ethics, underlining the legitimacy of who we claim to be as individuals, societies and even as a species. They pave the way for our behaviors and cultural variety. Whether dramatic, romantic or humorous, our stories are mind-provoking triggers that grant us a sense of belonging and identity. These, in turn, strengthen our context in a world that we are constantly trying to figure out and harness for the benefit of our lives.

We use our stories to engage others in all aspects of an experience that we decide is relevant, from information to feelings, discussions and conclusions. Through them, we enable comprehension and understanding, developing cognitive and emotional relationships between the story, the storytellers and the recipients. In short, our stories are the way we construct our lives around meaning. If homo sapiens appeared on the stage of this world some 300,000 years ago, then it is safe to say that we are living in a 300,000-year-old storyteller's festival.

All over the world, people live in accordance with the consequences of their story-making.

The red, yellow, blue and white colors stood out amongst all the greens and browns that surrounded us as we walked through one of the rich, beautiful forests in Madagascar. It was a misplaced and strange sight: Next to a water creek, one of the trees was wrapped in colorful ribbons (Photos 43 and 44, pp. 184–185). A young woman was facing the tree. Barefoot, her hands clasped together, her lips silently moved in what was undoubtedly a prayer. When she finished, she slowly took a few steps back while still facing the tree. She then turned to her shoes, quickly put them on and rushed away. Local *Malagashi*[8] explained that the tree is believed to have spiritual, godly powers, granting health, fertility and good fortune to believers.

In the Tibetan mountains, hanging colorful flags is a way to represent reading the Buddhist scripture. Called "the wind reads the scripture", the flags have a mission to carry scripture to the gods. Flags of five different colors are hung all around houses, poles and roads. Each color represents one of five different elements of life, symbolically representing that life is a circle: Red represents fire and symbolizes a warm family and brilliant career. Yellow represents the soil, symbolizing a rich and powerful life. Blue is water and sky, symbolizing boundless wisdom. White represents metal, symbolizing purity and eternity. Green represents the forests, symbolizing a long life. The more your scripture flags are blown by the wind, the more your prayers will be answered and your life will be better. Scripture and symbols are also placed on small sheets, called "Lungta" — "wind horse" in Chinese. Used in Tibetan ceremonies, the "wind horses" are tossed by believers in the wind, ensuring that the gods will receive and hear the prayers (Photos 45–47, pp. 185–186).

Probably the most widespread example of the impact our stories have on our lives is the development of religions. Briefly, religions are systems of faith-based stories that attempt to explain the world and our place within it, leading to commanding rules and behaviors that define the very practice of life.

---

[8] **Malagashi** — Natives of Madagascar.

Orthodox Jews believe that the land of Israel was given to them by a God who also ordered what and how to eat, bath and even reproduce. Every year, they will fast on certain days solely because they believe their God so commanded. Some regard biblical stories and timing literally, and will therefore insist that the world is less than 6,000 years old.

Similarly, some Christians believe that Christ walked on water and that Noah had an ark. Catholics have priests as intercessors. They must be baptized and confess their sins to priests who prescribe what they believe to be just penance. Catholic couples refrain from birth control and are unable to divorce. They, too, have fasting and certain practices called Lent, ordered by traditional stories that call for symbolic replication of Christ's sacrifices. In the early years of the 20th century, almost 2,000 years after her death, three children on a remote hill in the village of Fatima in Portugal repeatedly claimed they saw and spoke with Christ's mother, starting a story that led to a snowball of events that have since turned the city into a popular, and very prosperous, site of pilgrimage, where individuals who believe they have sinned can crawl on their knees and hence receive forgiveness (Photo 48, p. 187).

Observant Muslims believe that Muhammed ascended to the sky on a horse, commanded them to abstain from alcohol and pork, pray five times a day and fast for a month called Ramadan in commemoration of a meeting he had with an angel.

The list goes on, with every religion displaying a multitude of behaviors, rituals and habits, all a result of the stories they hold as truths.

Discussing religion as stemming from and connected to human stories is one of the most difficult discussions to have. Many followers of different religions will believe in them so much that they will passionately argue that even placing them in the same sentence with the idea that they are stories is misleading, not to say sacrilege and desecration, because their religion is "the truth", and definitely not "a story". Some will go as far as claiming that anything other than their religion is heresy. Not surprisingly, some religions are a rival of other religions; some religious groups will not settle with merely claiming the others are wrong, and will go as far as allowing and even encouraging murderous rivalry justified as "their gods' will".

When stories that are based on faith and belief define the intricate details of our lives, we are treading an area that leaves no room for argument. When our stories combine faith with distorted use of scientific knowledge or twisted facts and figures for so-called logical basis, such as the case of eugenics, Nazism, sexism, racism and more recently, vaccine-deniers, they become lethal.[9]

We are strongly committed to our stories because we need them. They help us make sense of the world, put things in order, make our lives richer and help us develop values, ethics and meaning. They fuel our imaginations, our creations, our decisions and actions, for better and for worse. Without them, we would still be roaming the savanna, indifferent to what the rest of the world has to offer.

Our stories will motivate us to bring our dreams to fruition, sometimes regardless of their accuracy. To this day, we have no idea if a leader named Moses ever existed, whether Galileo did indeed say, "And yet, it moves,"[10] or if an apple fell on Newton's head. It can reach a level of absurdity, as the case of a group of medical doctors who invested their time and energy in the research of James Bond, a complete figment of human imagination, so that they can conclude if his drinking habits were healthy — as if anything else about his abilities makes sense — and went as far as publishing the results in the *British Medical Journal*.

Reality proves that the impact of these stories isn't in their truthfulness. Their impact lay within the story itself. They inspire generations, teaching us about leadership qualities, the importance of observation and logic, strengthening the essence of qualities and meanings that are of importance to us. We cling to our stories so

---

[9] Science denial as well as science abuse will be further discussed in Chapter 6.

[10] Moses is a biblical leader. "And yet, it moves" is a saying attributed to Galileo Galilei after he was forced to retract from his claim that the earth revolves around the sun, thus declaring that scientific facts don't change even if some disagree and forcefully wish to ignore them. The legend of an apple falling on Newton's head is used to describe the importance of observing everyday phenomena as a source of logic and understanding.

stubbornly that we claim to have developed an entire scientific area as a result of the eccentric, not to say disturbing, behavior of another imaginary friend, declared by no less than the encyclopedia Britannica as the "Pioneer in Forensic Science": Sherlock Holmes. A man who never actually existed is declared to have pioneered an entire field of science that became the leading investigation paradigm, including methodologies, knowledge and understanding of modern-day police forces all over the world. Had the encyclopedia declared that Arthur Conan-Doyle, the genius author who created Holmes, was the pioneer of a whole new science, it would have been more accurate and appropriate — but also much less interesting, just like my history textbook.

So, Sherlock Holmes it is.

May the story be with us.[11]

Creating[12]

"Can we hear all the way to China?" asks young Ellie at the beginning of the 1997 movie, "'Contact"'.

"Sure, on a really-really clear day," her father answers.

"Could we talk to the moon?" she persists.

"If it's a big enough radio, I don't see why not."

"Could we talk to Jupiter? What about Saturn?" as her distances grow, the strength of the radio he offers grows accordingly.

"Are there people on other planets?" finally, her true dream spills out. With her, we will enjoy the rest of the movie, pursuing it until she finally creates the means she needs to receive an answer.

Story-making requires the ability to, well, make, in the most practical sense of the word. From the 75,000-year-old hunting story of the bushmen in Africa to this very day, our stories are so vivid that

---

[11] Paraphrase based on the popular sentence from the Star Wars movies, "May the force be with you", the Force being an omnipresence, a mysterious cosmic power for the good which guides and influences what happens and can bestow good luck.

[12] Creating something new involves creativity, design, modification of objects and so on. The act, in total, is an act of creating.

we end up not only believing them but turning them into man-made devices and disciplines. We confidently move through the process of imagining, designing and building new and ever-improved means to help us make our stories come true, a process that requires skills that are way beyond simple tool-making.

Any toddler presented with play dough or Lego blocks will use his or her imagination to create something. The toddler intuitively knows that the same tool can be used for different purposes in different contexts, real or imaginary, bringing innovative flexibility, adaptation and even fantasy into the equation. Moreover, we know how to build tools that work with each other as systemic machines, bringing complexity into the picture. Finally, we know how to build tools that dramatically expand and transform our biological abilities and environments, bringing design and constant change into the discussion.

Our curiosity drives us to use our tools in new, previously unexplored contexts, bringing open-minded discovery into our lives. It is this urge that sent Antonie van Leeuwenhoek to turn his magnifying glasses on a drop of water, literally discovering a whole new micro-dimension and world; that made Galileo turn his telescope to the sky, literally expanding the macro-horizons of our understanding; made Fleming poke into the character of the unwanted contamination on his experiment plates, forever changing our ability to treat and heal some of the most lethal diseases; the list goes on. The need to continuously explore, invent and create is what enabled us to develop transportation, medicine, agriculture, communication and much more, allowing us to travel beyond the capacity of our feet, see beyond the capacity of our eyes, talk beyond the capacity of our voices and live beyond the capacity of our unassisted bodies.

There is growing evidence, dating several millions of years ago, suggesting that the development of the human brain, our intelligent and complex ability to develop tools and the development of sophisticated communication and language are evolutionarily combined. Several millions of years allowed a step-by-step, spiral co-evolutionary process leading from our ancestral hominins to us. In an intertwined process, our hominin ancestors communicated to their peers how to make and use complex tools, further developing cognitive abilities leading to a growing diversity of even more sophisticated tool-making, communication, more cognitive development and so on.

This process expanded the biological ability of hominins to survive. Hunting and cooking devices enabled access to a wider range of foods, allowing the ability to process food more efficiently, making foods more edible and yielding more calories. It expanded their ability to defend themselves in a fight, flight, or freeze situation. It led to building temperature-controlled shelters. It enhanced relationships and interactions by creating and exchanging ornaments, gifts and most importantly, knowledge.

Taking these advantages into consideration, exceptional tool-making abilities is unmistakably one of the most important developments in the evolutionary history of hominins in general and homo sapiens in particular. It is so profound and unique to us that it deserves a more extensive look and will be discussed in Chapter 4. I am emphasizing here that practical making of any kind doesn't stand alone, as a quick look at Chimpanzees hunting bush babies with wooden spears reveals. Their primal ability to make ingenuous tools hasn't developed any further, for lack of the development of the other parts of the human system. Only when working in full combination with all the other components of "the big five", the interaction that is unprecedented in comparison to other species is produced.

Communication via Abstract Symbolic Systems

"No English, no English" was the answer I received from every person I desperately turned to for help. Lost in the streets of Athens, Greece, I was frantically searching for my hotel. I was 20 years old, on a trip away from Israel on my own for the first time, and Greek was not a language anyone thought to teach us at school. All I knew was the name of the hotel in English — The Apollo Hotel, in central Athens. With no one to understand me, I was lost.

Then, I saw a large sign with letters that looked familiar. I knew why: They were the letters we used in Math class. I started deciphering them. Alpha, Pi, Omega, Lambda, and knowing what the next two letters will be before I saw them — another Lambda and Omega — I shrieked with happiness and relief. The letters used in Math class, a necessary subject that I admit to have quite reluctantly studied at school, came to the rescue. It was the first time I was grateful for the hours I had slaved over equations, not because they helped me calculate, but because they helped me read. They spelled "Apollo".

I had just found my hotel.

We are *the* symbol-creators of the world. Be it a physical object we created, a new thought, new sounds, a new understanding, discovery or any other creation of ours, once we make something, we want to engrave it in stone, write it across the sky or at least publish it in a way for everyone to see, understand and admire how clever we are. The sense of achievement isn't fulfilled until we feed the urge to show and tell. To do that, we need to agree with each other on what and how we will represent what it is exactly that we want to express. Whether we begin with A–B–C, Do–Re–Mi or 1–2–3,[13] the idea is the same: We are using our highly developed and extremely unique *symbolic systems.*

We are the only organism that we know of on this planet that develops recognized and accepted symbolic systems. These systems enable us to communicate with each other at a level of sophistication that no other organism can, from explicit, practical information, such as where food and water can be found, to intricate and abstract ideas, such as time, thoughts and emotions. Our languages are examples of these symbolic systems, including spoken, signed and written codes. They are collections of representations, communicating meanings as agreed upon by a group of people. They have developed from carving and drawing accurate and descriptive figures, representing concrete thinking, to using no more than dots, lines and curves to symbolize meaning, expressing high-order abstract thinking skills. That's exactly what I am doing at this very moment: I am using a widely accepted system of symbols that you

[13] **"ABC, 123"** — the 1970 hit song performed by the American group "Jackson 5".

and I both decipher in the same way to communicate my ideas to you.

Our symbols are the cognitive, emotional and social tools we use to communicate meaning, reason, emotions, perception and instructions. In order to do so, they must be both recognized and understood by the users, who can be a group of people, a community, a society or humanity at large. We continuously develop symbols in the context of time and culture, enabling us to respond accordingly. For example, when we see the symbol X written on a road, it usually means a crossing over. On an entrance, it will probably mean no entry is allowed. When appearing on a text, the same X may mean that what we wrote is wrong, in an article about genetics and heredity it may represent the female chromosome, in other contexts it will mean the number 10.

The graphic symbol is the same.

The contextual interpretation of its practical use, very different.

Cultural behavior and development require communication within a context-associated symbolic world. The first time I travelled to Vietnam, I was shocked to see swastikas all around, proudly decorating places of worship, buildings, restaurants and gates. The explanation that it was an ancient symbol of prosperity and good fortune didn't change my feeling of contempt toward a symbol that represents, to me, the genocide of half of my people. Learning that the origin of the word "swastika" is Sanskrit and means "well-being" didn't help either. For me, it will eternally mean the worst evil humans can do. But, to the Vietnamese, Chinese, Japanese and other far-east countries, it is just an innocent religious symbol used to ornament their environment and bring them good luck (Photos 49 and 50, pp. 187–188). Judging by the words of the young man in Denmark who purchased a large bronze swastika and other Nazi symbols as WWII memorabilia at a local market in Copenhagen, he regarded it as a representation of the values of a regime that he wished would return (Photos 51 and 52, pp. 188–189).

The graphic symbol is the same.

The cultural context, and the thoughts and actions connected to it, extremely different.

Some of our symbolic systems are culture-independent, accepted and understood universally. Thus, anywhere in the world I may go, from Harvard in Boston to Weizmann in Israel, from the Academia Sinica in Taiwan to McGill University in Canada, and regardless of the language I may encounter — Mandarin, Hebrew, Greek or Portuguese — the specific drawing of these lines and curves will always and everywhere be recognized as symbolizing the molecule of water (see Fig. 1).

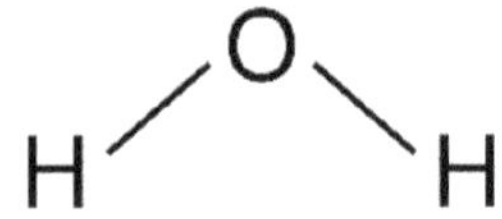

Fig. 1. H_2O molecule, commonly known as water.

Create a slight change in the arrangement of these lines and curves — as shown in Fig. 2 — and a completely different symbol appears.

Fig. 2. Boy, girl, you decide...

The lines and curves haven't changed. Our perception of their meaning has.

Almost 400 years after they were written, every child who takes piano lessons can learn how to turn these signs into the music that J.S. Bach intended them to be (see Fig. 3).

And almost 100 years after he died, completing only twenty-five percent of what was to be his greatest masterpiece, Gaudi's plans for the completion of the *Sagrada Familia* monumental church in Barcelona can be followed to the dot if we so desire, just because we

Fig. 3. Music sheet of Bach's Prelude.

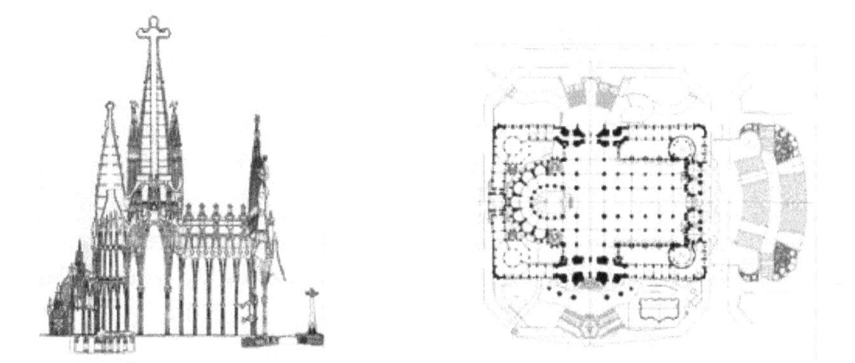

Fig. 4. Sagrada Familia plans.

can understand the blueprints he left describing how to do so (see Fig. 4).

The development of symbols as a means of sophisticated communication represents a delicate process of co-evolution and the emergence of combined cognitive, linguistic, motoric and other skills necessary to create, understand and spread abstract systems.

While evidence has been found suggesting that hominins from as far back as 120,000 years ago produced deliberate engravings associated with symbolic behavior, research regarding the exact timing of widespread development of symbolic communication is still to be continued. An array of carved objects, personal ornaments, musical instruments, art and other artifacts of concrete as well as abstract nature have been documented in many parts of Africa, Australia, Europe and Asia, as well as in association with Neanderthals. Most date between 30,000 and 40,000 years ago, though some date as far back as 70,000 years. Deciphering their purpose and the meanings they carry is still underway, but there is no doubt that these findings are evidence of the symbolic communication practiced by developed cultures. It is obvious that they are the result of thought, production and the desire to communicate content to other individuals. They are the means by which we keep our stories alive long after we are not.

Change

It was a dream come true. After years of biology studies, including dragging ourselves in the middle of the night to search for owls, hauling together up a mountain to see a rare Iris,[14] "schlepping" through caves to watch luminescent fish, my husband finally understood the hint and arranged the perfect birthday gift I could ask for: a trip to the Galapagos Islands. The origin of the "Origin of Species".[15]

I was standing at the top of the summit of Bartolome Island in the Galapagos, overlooking a breathtaking view of the volcanic Pinnacle Rock. Surrounded by various volcanic formations, tuff cones, lava flows and spatter all gleaming in deep black, red and

---

[14] **Iris Hagilboa** — a beautiful flower that blooms for a very short time on a specific ridge in Israel.

[15] ***On the Origin of Species***, the 1859 book by Charles Darwin that introduced the scientific theory of evolution, based among other things on observations and findings from his trip to the Galapagos Islands.

brown colors, what struck me most was the evidence of change and adaptation.

Almost barren, this is one of the youngest spots in the Galapagos. On our climb to the top, we spotted the Tiguilia shrub, a plant that is highly adapted to growing in the harsh, primordial environment that the island has to offer. It is considered a "pioneer plant", leading the way for others to follow, slowly inhabiting and changing the landscape. This process is known as primary succession: the development of a newly formed, primary habitat. A few spurge plants were scarcely scattered here and there as well, indicating that the process was well on its way.

Other, more mature islands in the archipelago have already had the time to develop more sophisticated vegetation, with many more species of plants and, consequently, rich wildlife, including insects, birds and reptiles (Photos 53–61, pp. 189–193).

Looking at the differences between the environments of the islands, it was as if I was looking at the nuts and bolts of a gigantic clockwork of time. Darwin made more sense than ever before. There was no doubt in my mind: I was standing within the very core of one of the most beautiful, exquisite definitions of change: *evolution.*

The one thing we can count on, stronger and more stable than anything we can think of or describe, the most constant character of the universe that is found in everything, everywhere, always, is change. Everything changes. Moving, vibrating, thumping, spreading, shrinking. Nothing stays still. Particles, atoms, molecules, organelles, organisms, environments, galaxies, the universe. Interaction of molecules leads to change in their properties. Growth and development are synonyms of change. Change is at the basis of ongoing migration of species, be it microorganisms in search of favorable temperature or light conditions, or insects, birds and mammals driven by the search for food and breeding sites, carried out regardless of high danger of predation and high costs of mortality, as revealed by the great migrations of wildebeests in the Masai Mara or the salmon fish across the Atlantic.

Change is our driving force, although we often have mixed feelings about it. When we explore our reaction to change, a strong difference can be seen when comparing our reaction to it as individuals with how we deal with change as a species.

As *individuals*, more times than less, we tend to resist change. We find it difficult to deal with the uncertainty and insecurity that comes with changing places, jobs, habits and social environments. Change means departing from what we are accustomed to, accompanied by a sense of loss and awkwardness toward the different, new and unknown. Leaving the comfort zone of habit can be, well, uncomfortable. Sometimes, extremely so. Not something most individuals will rush in to.

But, as a *species*, homo sapiens exhibit exactly the opposite. We *adore* change. We are never satisfied, can never find the very best conditions of shelter, safety, food supply or health, for one very simple reason that is embedded in our nature: We are always, but always, searching for more. We constantly crave for what we don't have, and are willing to act to quench our endless thirst. If we just change where, what and who we are, all our dreams will come true. We never live happily ever after, simply because we are always certain that something better lay ahead and our curiosity to discover what it is and then possess it is insatiable. Somewhere over the rainbow, dreams that we dare to dream really do come true[16]; if I were a rich man, I'd have a staircase leading up and one leading down and one more leading nowhere just for show[17]; if we cross the ocean, we will discover something different and worth discovering. No wonder one of the most successful, long-running plays off-Broadway was a musical titled "I love you, you're perfect, now change".[18]

We are an extremely curious and ambitious species, always designing and changing our environments, others and ourselves. Whenever we accomplish or acquire something, we look for ways to

---

[16] **"Somewhere Over the Rainbow"** — song from the 1939 film "The Wizard of Oz", sung by American actress Judy Garland.

[17] **"If I were a Rich Man"** — song from the 1964 musical "Fiddler on the Roof".

[18] **"I love you, you're perfect, now change"** was the longest running musical comedy on an off-Broadway stage, performed more than 5000 times from 1996 to 2008.

improve and fit it into our ever-developing new whims. It is one of our inner driving forces that, in combination with the other four, has taken us from huts to houses, from wagons to spaceships, enabling us to more than double our longevity and eradicate some of the worst diseases we suffered from. We are in constant pursuit of *something better*, be it happiness, wealth, health. We are driven by the notion that we can and should change.

Acknowledging change as a typical characteristic of homo sapiens, some argue that it is because we are the only species who is constantly trying to defy death. The ultimate, finite change of all, we refuse to accept the banality that mortality casts on our very existence. So, we hang on to the belief that there must be meaning and a purpose to our lives, realizing how short they are and always striving for more time before it's all over.

There is a large body of research showing that having purpose and meaning in life increases overall well-being, a sense of satisfaction, improving mental and physical health and resilience. Trying to control our lives, we continuously create means to improve and lengthen them, adopt behaviors that lead to a healthier and more prosperous existence, inventing ways to leave a mark and be remembered, desperately trying to change the fact that we will one day die and eventually be forgotten, undoubtedly our greatest fear of all.

If eternal life is the goal, the best way to achieve it isn't by having children. It's by the power of a strong story, the kind that creates change that is so impactful that it survives and thrives long after we are gone. They who create a story that changes the lives of others shall be remembered. Whether that change is for better or for worse is completely up to us, as will be discussed in the following chapters. The story of humans is a network of narratives filled with names of individuals who are remembered because of the changes they have inflicted on our lives. The impact of their stories lives on.

Back to Bartolome. On the way down from the summit and back to the shore, my eyes suddenly caught the color yellow: Amidst a Tiguilia shrub, almost as if it were growing from the same roots, a single Shore Petunia was blooming.

Change wasn't on its way. It was constantly there.

Inter-generational Transfer

He lived some 300 years ago and to this very day Sir Isaac Newton is considered one of the leading scientists of all times, who transformed the way we understand and think about the universe. His contribution to physics was accepted as the most important and influential work of modern science.

Yet, had he returned today to his alma matter, Trinity College in Cambridge, he wouldn't recognize some of the most leading, core theories in physics, simply because since he has gone, others physicists have followed and further developed the field.

Newton was regarded by many of his peers as an unparalleled genius, so much so that every research and discovery that followed was expected to "fit in" with his theories and laws. Poet Alexander Pope surpassed this approach in his epitaph to Newton:

> *Nature and nature's laws lay hid in night;*
> *God said "Let Newton be" and all was light.*

During the 19th century, scientists began to reluctantly notice that not all was well in the Newtonian kingdom of physics. Evidence was beginning to show that there were phenomena that couldn't be explained by the accepted universal laws. According to Newtonian physics, time, space and geometry are absolute and unchanging. Then, along came new discoveries in electromagnetism and mathematics. These eventually led at the beginning of the 20th century to the theory proposed by the "new and improved" scientist in the neighborhood, Albert Einstein, who realized that contrary to Newton's laws, time, space and geometry are not absolute but relative and depend on the physical environment.

New theories were developed, others followed, more continue. The physics of the universe are continuously being researched and discovered, far from being unraveled, as are other areas researched by humans.

We are the *perpetuum mobile*[19] species, forever on the move of development. Generation after generation, we rely on what has already been revealed, continuing to evolve from the point that the previous generation left on. This is very much contrary to other organisms, such as the African weavers, who generation after generation have been weaving the exact same kind of nests, or the leaf-cutter ants, who have continued to grow the same fungus in the exact same way for millions of years.

The offspring of one generation of homo sapiens receives knowledge and know-how from its parents and then stubbornly continues to introduce changes, developments and improvements, without the necessity to have each generation begin to learn everything all over again. We don't need to search for stones that can be knocked together to create fire, like our ancestors did. All we need is a match that we bought at a local store. We don't use papyrus to write on anymore, we hardly even use paper, we use some buttons on a keyboard. Most likely, our children or their children will use something else, yet to be developed. As stated by Newton himself, "'If I have seen further, it is by standing on the shoulders of giants."'

Newton is far from being the only "human giant" who if returned today would find difficulty in understanding the very expertise that awarded him this title. Moreover, it no longer takes 300 years to create substantial development and change in any given area. Our extraordinary ability to transfer and develop knowledge from one generation to the next stands out in exceptional ways when observing the 20th and 21st centuries.

For collective reasons that will be discussed in Chapter 4, in less than 200 years, human abilities in all aspects of life have known an unprecedented pace and race of transformation. Medicine, com-

---

[19] **Perpetuum Mobile** — a hypothetical everlasting motion of bodies or machines that continue to function and work forever without the input of external energy, which is impossible according to the laws of physics.

munication and transportation are but few examples of areas that would be unrecognizable, even bizarre, to anyone who lived a mere century ago, or even less.

As a high school student, I would search for information in a large building housing a library, containing endless shelves lined with alphabetically organized books and journals, finding my way using an elaborate index and the help of a librarian.

Today, my grandchildren carry their libraries in the palms of their hands. All they need to do to find something is use a key word in a search engine that will find and offer a variety of options related to what they are looking for.

When my children travel to other continents of the world, they send WhatsApp messages that arrive instantly, a fact of life that is so trivial today that we can't imagine what life looks like without the satisfaction of constantly knowing exactly what's going on and what it looks like, all in real time.

And, while the Wright brothers achieved the first successful flight in a heavier-than-air powered aircraft in mid-December 1903, lasting only 12 seconds, it took less than 66 years, within less than one person's lifetime, for the first human being to land on the moon.

And, we're on our way to Mars.

***

The unique combination of the "big five" is the fuel that turns the spiral circles of our existence. A unique, vibrant and ongoing human system, it is greater and more powerful than the sum of its parts. It is the launch pad from which four major aspects takeoff, altogether part of our special identity. These four grind and grow through a distinctive relationship at a level of complicated development found in homo sapiens alone:

(1) The manifestation of sophisticated *intelligence* in general;
(2) *Technological intelligence* in particular, leading to exceptional technological competencies;
(3) Our inclination to not only teach and learn but to *educate*; and

(4) Most recently developed, the paradigms and practicalities of *modern science.*

The next four chapters will discuss each of these aspects, their inseparable interactions, how they work with the big five and the impact all this has on the story of our lives.

Chapter 3
It's Complicated*

The animal community living on the island of South Georgia was watching us with curiosity as we landed. It was cold, but my heavy yellow parka and high waterproof boots kept me warm. The shore stretching in front of me was the breathtaking sight of a remarkable number of lion seals, king penguins and elephant seals. Their gaze accompanied us as we disembarked our zodiacs, some staying aloof, some ignoring us completely and others watching attentively.

Wondering what these strange, yellow guests — namely, us — were up to, some of the animals boldly started coming up close to explore. It was obvious that they were the masters of this environment, checking out the newcomers. As they pondered and poked our belongings, backpacks and cameras included, it was also clear that they were curious about us at least as much as we were about them (Photos 62–65, pp. 194–195). Some called to others as if telling them to come and see, some quickly lost interest and returned to their own business, playing with each other in what looked like a graceful encounter despite their heavy and seemingly clumsy bodies.

---

* **It's complicated** — a 2009 American film by Nancy Meyers, starring Meryl Streep, Alec Baldwin and Steve Martin. The film explores complex relationships between variables — in this case, people.

I was very busy talking to a king penguin when I noticed a small group of them waggling their way in a neat row toward me as they wanted to approach the sea. Like in a cartoon, they almost bumped into each other and abruptly came to a stop when they realized I was standing in their way.

Amused by the obvious confused expression on their faces, I didn't move, waiting to see what they would do. They turned to each other, then to me, then again to each other. There was some flopping of wings and gestures of beaks, including looks that were sent in my direction with an unspoken question, "What is she up to? Is she going to get out of our way so we can continue?"

Refusing to move, I was eager to see how they would solve the problem.

Finally, the first penguin had a bright idea: He nodded to the others, then turned left and created a careful detour around me, as they followed (Photos 66–72, pp. 196–199).

There was no doubt in my mind: These are creatures of intelligence, displaying explicit reasoning, information processing, communication, understanding, decision making and simple problem solving.

"Man is the only animal for whom his own existence is a problem he has to solve," said German philosopher Erich Fromm, addressing some of the most ancient questions we ask ourselves: Who am I, how did I get here and what is the purpose of my life? One of the main driving forces of these ongoing questions is our fascination with the similarities and differences between ourselves and other animals, our ongoing curiosity and natural leaning toward relativism and comparison, striving to understand the living world in terms of our own abilities.

As discussed in Chapter 1, it is widely accepted that components of intelligence, such as communication, simple signal transfer, mourning, deception, ritual and cultural development are found in other organisms. Great apes, our own family members and closest relatives (Photo 73, pp. 199), show both cognitive and emotional behaviors — manipulation, tool use, hierarchical behaviors, use of simple symbols, numerical concepts and possibly displaced reference.[1]

However, these yet to be proven abilities are far from being as complex or sophisticated as can be found in humans, and can be considered as "coming soon advertisements" of what is to be further developed in humans: The emergence of properties that interlocked and co-evolved into a coherent, powerful and versatile ***human intelligence system.***

Anything You Can Do, I Can Do Better[2] — or Not

"Put your pencils down, *now*!" Michelle and I, only in fifth grade, were standing outside the door, eavesdropping, when the supervisor's sharp voice thundered its way to us. Michelle's sister, Tami, was 3 years our senior and inside that classroom. It was one of the most important days at school: The eighth-grade placement exams, determining the participant's IQ, which meant the level of your intelligence.

In our words, the exam decided whether, in the eyes of the world, you are smart.

Or not.

Tami, Michelle's sister, came out of the classroom, sobbing. She was certain she had failed the exam. We all knew that the placement exams controlled our future: A high score meant you were on the right track to getting into better schools, being offered better opportunities of higher education and eventually leading prosperous, better lives. A low score meant being sent to a vocational school, or in other words — it meant you were second best, "plain", probably on your way to be part of the blue-collar working class, a way to say

---

[1] **Displaced reference** — the ability to refer to events that happened in the past or will happen in the future. Orangutan females are thought to be able to communicate past events to their young.

[2] **Anything you can do I can do better** — a song from the 1946 Broadway musical "Annie get your gun", composed by Irving Berlin. The song is a duet between a male and female, attempting to outdo each other in increasingly complex tasks.

you probably weren't intelligent enough to someday be accepted to a university and be part of the "elite". We had no idea what exactly "intelligence" meant, or even what the "better lives" it led to were or what being part of "the elite" meant, but we knew it was extremely important.

Ever since the third-grade selection exams had differentiated us into three levels by what was considered our "intelligence score", we had all been anxiously working hard to be accepted to what was called "group A", the group of "gifted" kids, "gifted" because they succeeded in math, science and language. Gifted kids were precious. They were the future. We all wanted to be precious. It smelled of success, prosperity and happiness.

Watching Tami cry, I couldn't understand what went wrong. She was a bright girl, brilliant violin player, who excelled at telling stories that combined deep knowledge in history and literature. She was funny, warm and outgoing. We all loved her. The exam, however, was a 3-hour torture of deciphering logical sentences, Math riddles and psychometric shapes. Two weeks later, when her results came in, her smile returned as her scores were high. Her future was saved. I was too young to explain to myself why, but I was certain that this one-tracked determining process was completely wrong. Something wasn't working. We had to be more than the sum of our performance in math, language and some shapes on paper. Our intelligence had to be more than just the score we achieved in a test on a specific day.

It had to be more complicated than that.

Somehow, I fit in to the system's standardization requirements, but I witnessed too many peers who didn't. My uneasiness grew as years passed, and the gap between success in those placement exams or at school in general and success in life became even stronger. Naomi, who was usually absent from class and an average student at best, became a wonderful ballet dancer. Dan, who was a trouble-maker that nowadays would probably be diagnosed with ADHD, became a successful global businessman. Tom, who flunked most of the classes, became a brilliant orthopedic surgeon. And, Dana, the brightest student of us all, hung her high school diploma on the wall and decided that it was enough for her.

The tests we took to differentiate our intelligence level — the later infamous IQ exams — had nothing to do with who we really were or who we were to become. I couldn't help but feel that the fact I was successful in a rigid education system, passing the necessary exams and floating from one degree to the next, was no more than a matter of sheer luck.

After I received my PhD in life sciences and began my career in science education, an accumulation of experiences changed my perception of human intelligence and its connection to the story of humanity. Two major pillars served this process: **The general systems theory** and theories of **intelligence**.

To understand the connection between the general systems theory and intelligence, we must first pause to understand what led to the development of the theory, its leading principles and how they are expressed in everyday life.

The System Dance[3]

I was fascinated by the poem in front of me. It described a dichotomy: On one side was the harsh, technical, reduced point of view of each one of the parts of a rifle that must be named and memorized. On the other side were the sensuous, flowing interactions revealed by nature as a holistic phenomenon, whose parts cannot be considered separately. Henry Reed's "Naming of Parts"[4] was the perfect metaphor for what I felt as I rapidly approached the end of my graduate studies.

Diving deeper and deeper into the technical, operative details of a specific gene, I felt like a horse with eye blinders, made to ensure

---

[3] Paraphrase of **The Skeleton Dance**, based on the spiritual song **Dem Bones** transformed to a children's song, teaching them the connections between different bones necessary to create the full functioning skeleton.

[4] **Henry Reed** (1914–1986) was a British poet and journalist. His poem "**Naming of Parts**" was part of his most famous work, "Lessons of the War".

that my attention is focused only on a small portion of what nature has to offer and science can discover. The irony of the poem helped clear my thoughts. Too often, we have separated our sight from the bigger picture of nature's dynamics, disturbances and points of balance. If we continue, it might soon be lost to us forever.

When we explore the world, ourselves included, it is obvious that our reality is made of an endless variety of components, invariably interacting with each other. Endlessly curious beings, one of the best ways we have developed to make sense of this reality is science: A specific school of thought and methodologies aimed at building and organizing the world, based on accumulated evidence, testable explanations, logical critique and so on.

The complexity of the world poses challenges to our efforts to reveal and process the growing information we gather regarding its components and the ways they work and interact. To cope with these challenges, many times we separate the parts, reducing our exploration to bits and pieces of the bigger picture.

This is what we do when, for example, we explore the interaction between a certain protein as it connects to a specific neuronal cell, when both are in an artificial cell culture — what is known as in vitro ("in the glass") experiments. Whatever happens is true for the specific setting of the experiment, not necessarily for what happens in the same cell when encountering the same protein in the whole, living organism, with so many other factors surrounding them that may have substantial impact on the interaction — what is known as in vivo ("within the living") experiments.

This separation is part of what is known as the *reductionist approach*. It is based on the assumption that we can explore each piece one by one, understand what it stands for, then put the pieces together again to create the "bigger picture", just like we do when we try to put together a puzzle. It led to the creation of disciplines, such as physics, chemistry and biology, a method we developed as a necessity to understand and manage our lives. As knowledge and needs grew, more disciplines arose, sometimes by combining two or more, such as bio-chemistry, genetic-engineering and so on.

But, just as all the king's horses and all the king's men couldn't put Humpty[5] together again, it isn't so easy when trying to regroup the pieces we need to understand the world we live in, either. While it's important to explore the different components that constitute a whole, it's obviously not sufficient enough to truly understand that whole for one simple reason: Reality isn't based on separation. On the contrary, it's based on complex interacting connections. Without regarding these interactions, solving problems becomes impaired. It's like trying to address the needs of a child without regarding the parents, the school and the environment in which the child interacts.

The idea that nature, including humans, works as a holistic, complex system, has been part of human philosophy for centuries. From Thales[6] to Humboldt[7] and beyond, philosophers have described the relationships between parts of the cosmos as a large, interacting entity, made in itself of connected systems that influence each other through inseparable Gordian knots.

As the 19th century turned to the 20th, the impact of our scientific and technological abilities grew. Their growing use by industrial, political and social forces and their connection to global processes became deeper and stronger. The resulting complexity unraveled, and the level of sophistication necessary to cope with the challenges we face became widely discussed. Issues such as food security, energy

---

[5] **Humpty Dumpty**, a character in an English nursery rhymes, known for "having a great fall" only to break and never be able to be whole again, despite the efforts and good intentions of many noble men.

[6] **Thales of Miletus** (540 BC) is regarded as the first philosopher in the Greek tradition to have entertained and engaged in scientific philosophy. Breaking from Mythological explanations, he was the first to explain the world and the universe by naturalistic theories and hypotheses, modeling the path for philosophers who followed as a herald to modern science.

[7] **Alexander Von Humboldt** (1769–1859) was a German scientist, who pursued a unified approach toward diverse branches of scientific knowledge and culture, viewed by one holistic perception of the universe, laying the foundations of Ecology. He was also the first to describe human-induced climate change, laying the foundations of environmentalism.

and material resources, pollution, health and well-being, climate change and so on made it clear that more holistic approaches are needed if we are to understand and deal with them responsibly. We needed paradigms that consider the phenomena we explore as *systems*, interacting both intrinsically and between each other in various, sometimes unique, ways.

This realization led to the development of **The General Systems Theory**[8]: An approach focused not on the mere understanding of separate components but focusing on the synergy between them as well. The general systems theory formally defined that while attempting to understand the role of each component is important and even a must, it is no less crucial to understand the activity of the whole system and its connections to the environment. If we wish to truly be able to design our lives and environments, to continue developing our science and technology in relevant and impactful ways, we must explore the world through the prism of systems and their features.

So, the world is made of systems. What does that actually mean and why is it relevant in a discussion about intelligence?

A system is any group of parts, interacting with each other in a specific, highly regulated manner, creating a distinctive whole that is always larger than the sum of its parts. This means that the operation of a system cannot be achieved by each part working alone.

Moreover, the system itself operates in a holistic manner, meaning that if any given part is omitted, the entire system is influenced, sometimes lost. This leads to the understanding that the properties of a system are present only when the parts work as such. A direct example is the parts of a car: A wheel, an engine, a frame, seats, and so on. Only when all these parts are assembled and work together in a very specific way does the concept "car" awaken and have meaning.

As we look at the main characteristics of a system, we'll use the example of one of the central, leading systems that is meaningful for us all: Our health.

---

[8] **The General Systems Theory** was initially presented by Austrian Biologist Ludwig Von Bertalanffy in the 1940s.

Every system is characterized by several features:

- **Interactions within the system:** The operation of each part is influenced by the operation of the other parts.

  It is widely accepted that our health is a complex system, comprising several parts: Physical, mental, social, environmental and transcendental. Our health depends on the successful function of each part, as well as on the dynamic relations between them.

  We are all aware of the importance of dealing with a person's emotional, social and environmental conditions and not merely their physical status when we wish to cure diseases and promote health. There is accumulating evidence suggesting the influence of patients' expectations on the efficiency of medical treatment and its outcomes. Simply put, an optimistic, socially involved person, who also strongly believes she or he will overcome a certain disease, will probably have higher odds of overcoming the disease, when compared to a person with the same disease who is pessimistic, socially detached and is convinced that the disease is stronger than she or he is.

- **Boundaries:** When we decide which parts are included in the system, we are setting its boundaries and with them, our points of reference.

  Let's say, for example, that we want to understand the influence of air on our health. We can decide to discuss in detail the parts of our respiratory system, and settle with that.

  We can also decide to expand the boundaries and include other body systems, such as the blood, digestive and nervous systems, and our emotions when we breathe fresh or foul air, or when we use air to create music, blow soap bubbles or simply enjoy a balloon.

  We can further expand the system boundaries to include parts of the environment that influence the air quality in our surroundings, such as habits of other family members who may be smokers, or the level of polluting transportation and factories, maybe a nearby forest or other "green areas", and so on, all affecting the quality of air we breathe.

  Hence, each system is also part of a larger system.

- **Complexity:** The definition of boundaries also reflects the level of organizational order, hierarchies and growing complexity.

  Every single living cell is a system defined by the cell membrane. It will input nutrients from and output waste to its environment, necessary for its ongoing metabolism. It will do so whether the cell is the entire organism living in a pond or if that cell is one of many cells of a multicellular organism and its environment is other cells.

  This leads us to a higher level of complexity that can be seen in, for example, the brain. Also a system, our brain is made of many cells that operate together as an organ that can be such only via the specific multicellular brain structure. Expand the boundaries and you will encounter a much more complex system, including millions of cells of yet other organs, all working together to create an entire organism, such as a tree, a fish or a human being.

  As we define the borders to include the parts of the system we wish to consider, we also define its level of complexity.

- **Interactions with the environment:** Systems operate in an environmental context, communicating with it through inputs and outputs of matter, energy and information.

  For example, think of the input of food, its conversion to energy and matter in the body, the discharge of heat and excretion of waste; or the input of information, such as the influence of smoke and nicotine on our lungs and hearts, hopefully resulting in the output of behavior such as giving up smoking.

- **Control and feedback:** While by definition all systems are open to their environments, they will be so at different degrees.

  The brain, for example, is a relatively closed system when compared to the lungs or the digestive tract. This leads to different degrees of self-regulation and control of inputs and outputs *vis-à-vis* environmental changes, necessary for the ability of the system to react successfully to these changes. It requires ongoing, everlasting interactions of communication, feedback and regulation between the system parts, and between them and the environment.

Think about climbing a mountain, a high one. As the level of oxygen decreases, our body responds. Our breathing becomes faster, our head starts to ache, we lose energy, feel weak, nauseous, sometimes we might even vomit or worse. That's the regulation of our body, one of the most beautiful complex systems in the world, kicking in to alert us that there isn't enough oxygen to continue operation and that it is crucial to take care of the situation before it might be too late.

- **Dynamic change:** As discussed in Chapter 2, ongoing change is probably the most constant feature of the universe. Naturally dynamic by definition, systems face continuous fluctuations of conditions. The challenge is to successfully balance between these changes and the natural tendency to retain stability, while continuing to develop and evolve.[9]

  In living systems, us included, these are the mechanisms that work to sustain order under both expected and unexpected changes, be they welcome or disturbing.

  Think of it like this: A sudden accident has left us without a leg. Recovery depends on our ability to emotionally accept the loss, cognitively find alternatives, overcome physical pain, recreate former abilities as well as develop new ones, harness social support and acceptance, and continue with life.

When the dynamics of a system, living, environmental or artificial, react successfully to change, the system can be considered a healthy one.

The general systems theory is an elegant and useful tool when approaching the complex, dynamic phenomena of the world. It allows us to understand the world as an organization of interacting systems, diverse in size, structure and dynamics. While the realities we live in, design and change are sometimes challenging even for our own understanding to grasp, we can safely agree that the universe, our planet, environments, organisms, individuals, social structures and so on are all systems.

---

[9] Balance between changes to retain stability is known as **homeostasis**; doing so while continuing to develop and evolve is known as **homeorhesis**.

So is **human intelligence**: An ultra-complex and unique system of inherent, interacting capabilities that enable us to plan, create, solve problems and adapt to change in flexible, socially accepted ways.

Now that we know what a system is, we can discuss intelligence.

How Shall We Describe Thee?[10]

We claim to be the only living being conducting itself through the power of high-order, sophisticated intelligence. Obviously, the strength of human cognition is unprecedented when compared to other primates. Our intelligence has taken us far beyond what is biologically required to merely live and reproduce, a state we are so consciously aware of that we proudly call ourselves "Homo sapiens", literally meaning "the wise hominid", no less.

However, while we feel we know what intelligence means, it is one of the most difficult terms to define and trace the source of development for. In spite of being thoroughly researched, described and even seemingly measured, intelligence seems to be one of those ambiguous, even illusive concepts, like energy, time or love: While we claim to recognize it and are mostly certain that we understand what we're talking about, it's almost impossible to define accurately.

And, it's not for the lack of trying.

A society's development is based on the competencies of its human capital. The more developed these competencies are, the more beneficial it will be for the entire society. High intelligence skills are considered a key to the development of an intellectually advanced, skillful and successful culture. At the turn of the 20th century, Europe and the US were in search of ways to increase their ability to predict the cognitive abilities of individuals, especially children, who could then be nurtured and developed to become involved, meaningful citizens and, more importantly, to become accelerators of their country's desires.

To solve this problem, the French ministry of education commissioned in the early 1900s psychologist Alfred Binet to develop a test

---

[10] Paraphrase of **Elizabeth Barrett Browning**'s poem, "**How shall I love thee**".

that could help determine the intelligence of children so that those who are considered behind in their schoolwork could be "remedied", while those who are most successful can be "accelerated". The result was the Binet-Simon *IQ tests: Intelligence Quotient*, a score you received after completing a set of standardized tests.

Binet himself stressed the limitations of the test. He suggested that intelligence is far too broad a concept to quantify with a single score, insisting that it is influenced by many factors, such as emotions and social context. He further claimed that it changes over time, and can only be compared in children with similar backgrounds.

Nevertheless, it was the best tool available and rapidly became the basis for continuous development of intelligence tests. Institutions around the world, such as education systems, universities, military, industry, police and more, implemented screening and admission requirements based on the scores that applicants received on their IQ tests. They were used to determine job classifications and professional positions that a person was most suitable for. Education systems used them to screen and differentiate between "talented" students, therefore worthy of extra investment and nurturing to promise their fruitful futures, and "special needs" students, who require either additional intervention to stay in the "normative track" or to be sent to different academic environments altogether.

The rapidly spreading use of IQ tests was used to support the argument that intelligence is determined by biology. Biology determinists hurried to harness the tests to their side, claiming that they present proof that biological diversity is a result of inherited differences, determined at birth. Using scientifically distorted arguments, determinists claimed that socio-economic differences between human groups are of genetic origin. During many of the events of the 20th century, IQ tests played a meaningful part in support of eugenics — beliefs and practices aimed at "human improvement based on genetic quality". In addition to being used as the justification for the racial policy in Nazi Germany, the use of IQ tests played a role in the eugenics movement in the United States, leading to the 1927 Supreme Court legalization supporting forced sterilization of anyone who threatens to "dilute" genetic excellence. With a vision of eliminating "undesirable human traits" while endorsing selective breeding for what was considered positive traits, thousands of people were forcefully or deceivingly sterilized[11] under the premise of being "feeble minded" for achieving low IQ scores. Compulsory sterilization based, among other reasons, on low IQ scores, continued formally until the mid-1970s, when lawsuits and public awareness finally led to their halt.

As studies in psychology, education and sciences developed, there was growing agreement that basing the concept of intelligence on IQ test scores alone was insufficient. It neglects crucial aspects of mental ability, knowledge types and skills, all of which produce success in human society, such as creativity and social skills. Critique and frustration grew from what is described in the words of Albert Einstein: "If you judge a fish by its ability to climb a tree, it will live its whole life believing that it is stupid."

As the second half of the 20th century unfolded, broader expressions of intelligence began to be discussed. New theories and definitions of intelligence[12] were placed on the storyboard of homo sapiens, definitions that regarded intelligence not only as a set of cognitive skills but as *interacting, complex systems of competencies.*

Furthermore, it was also suggested that we have more than one single type of intelligence, and that we can develop a wide array of knowledge, skills and flexible performances in different domains as we grow and develop. These ideas defied the traditional paradigm that intelligence was all about cognitive abilities, adding emotional, creative and social abilities to the picture. It became more and more evident that the variety of expressions displayed by human competencies is a matter of *combinations*, and that these combinations are plentiful. Just like the six letter language of DNA and RNA

---

[11] The **eugenics** movement in the US is described, among other places, in the 2014 novel "**Necessary Lies**" by Diane Chamberlain.

[12] **A variety as well as new theories of intelligence** can be found in the bibliography: The theory of multiple Intelligences (Gardner, H); The Triarchic theory of Intelligence (Sternberg, R); The Theory of Emotional Intelligence (Goleman, D.); and Adaptive Intelligence (Sternberg, R).

creates an endless rainbow of genetic combinations, or like some twenty amino acids create endless combinations of proteins, the combinatorial mosaic created by potential human capabilities is striking.

In addition to the realization that intelligence is not about cognition alone, evidence continues to accumulate in support of the understanding that it's also not about biology alone. Fossils and comparative DNA research continue to suggest that all humans, in all continents, inherited our current brain structure and genome from our ancestors around 300,000 years ago. That means that the *hardware* for building the pyramids, writing symphonies or contemplating whether to be or not to be[13] was already "uploaded" long before it was actually expressed.

These findings join accumulating questions regarding the claim that the evolution of human intelligence is in correlation with the evolution of the size of the brain. Researchers have argued that the ability to cope with complex environmental challenges, such as the need to find food, avoid predators, perform within social groups and even successfully raise offspring, created the conditions favoring larger, more developed brains.

For example, I have already mentioned that mammal babies are born "half baked" — they cannot survive on their own, requiring close parental care before they can "leave home" and be independent. Research has shown that early childcare may be one of the driving factors of brain evolution through *positive feedback*: To accommodate birth with a large brain, human babies must be born early. Early birth led to the need for more care. More care requires more intelligence to be successful. So, the loop goes on, highlighted in humans more than in any other primate.

However, our brains aren't the biggest in the neighborhood. They're not even the largest when compared to body size. Research has also shown that the ratio between body and brain size in a variety

---

[13]**To be or not to be** — from Shakespeare's play "**Hamlet**". It is the opening phrase of the self-addressing monologue by Hamlet, dealing with profound existential questions.

of organisms isn't necessarily connected to intelligence, and can be a result of changes in body size that have no influence on the brain.

For example, seals are intelligent mammals, although their brains are relatively smaller than would be expected for organisms with their body size (Photo 74, p. 200). It is proposed that throughout evolution, low water temperatures led to natural selection of larger bodies, without affecting brain size. Thus, intelligence isn't necessarily determined by brain size, and the connection between the two is more complicated than previously thought.

One of the main questions still explored is a temporal one. While homo sapiens appeared at least 300,000 years ago, evidence of the increasing development of intelligent products such as sophisticated symbolic language, art and tools is no more than 100,000 years old, creating a meaningful and questionable time gap between the development of our brain and the development of the practices it enables. This gap is even more puzzling when considering that all cultures of humanity share similar behaviors of all big five — imaginative storytelling, artifact development, creativity, abstract symbolism and communication, and inter-generational education — regardless of geography. In other words, why haven't advanced literature, microscopes, satellites or philharmonic orchestras developed much sooner? And, what happened some 100,000 years ago that changed the situation?

These and many other questions are yet to be answered. While we know quite a bit about the *hardware* of our brain, i.e., the parts that comprise it, we are only just beginning to understand the *software* of our brain, it's onset and functioning, i.e., how the different parts function and interact under different circumstances. If we agree that the potential to cry "Eureka"[14] or land on the moon has

[14] **Eureka**!, meaning "I have found it!" in Greek, is the exclamation used to celebrate a new discovery or invention. It is based on the story of Greek scholar and mathematician Archimedes, who upon noticing that the water level in the bath rose when he stepped into it, suddenly understood that the volume of water displaced is equal to the volume of the part of his body he had submerged. He was so excited to share his understanding that he reportedly leaped out of the bath and ran through the streets, naked, exclaiming "Eureka! Eureka!"

been available to us ever since the dawn of mankind, then the questions of when and why we started using these abilities, enhancing them to work together, are still under exploration, exciting our imaginations and waiting to be revealed. One thing is obvious: The range of our potential abilities expresses itself in different ways in different individuals and in different contexts. They develop in correlation with a society's preferences and development, or in other words via *cultural connected evolution,* as the following example suggests.

I was walking with a group of six nature lovers from Europe through the oasis that lay between the canals and riverbeds of the Okavango Delta in Botswana, searching for whatever nature would offer. It was early morning, and the smell of fire filled the air. Deliberately set by local bushmen, fire is used to manage the land, returning nutrients to the soil and clearing the ground of unwanted plants — a highly disputed, primitive method that has been in practice since the beginning of agriculture, some 10,000 years ago (Photos 75 and 76, pp. 200–201).

After enjoying the sight of a group of impalas and one majestic great kudu that passed us, we approached a large, just-burnt field. Here and there, sparks were still quietly sizzling, as if waiting for an opportunity to rekindle, refusing to fade out. On the far side of the field, between green trees that were far from the reach of the controlled, now almost extinguished, fire, we spotted some elephants.

Eager to get closer, we noticed a wide, open path stretching on our left directly toward them. Intuitively, we hurried to turn and walk on the path around the ashes and sparks to avoid the smoke and heat as much as we could when Kendru, our local guide, abruptly stopped us. His voice was sharp as he scolded us, "Only *Faranji*[15] walk without using their brains."

"Aren't we going to see the elephants?" someone asked timidly.

---

[15]The name **Faranji** was coined based on the Persian term Farangi, meaning "Germanic", of "Franc" origin. It is still used throughout Africa and Asia mostly by locals, meaning "foreigners", "Europeans" or — as in British India and Ethiopia — "White people who travel", i.e., accurately describing the situation we were in.

"Of course we are," he replied. "But not like that! Unless," he added, cynically, "you want them to use you as a boxing bag!" and rolling his eyes, he repeated, "*Faranjis...*"

We waited, patiently. Kendru sniffed the air and set his gaze on the large field that had just been burnt. Then, to our surprise, he started walking directly through the center of the still warm ashes and glowing embers. He motioned us to follow.

We looked at each other, hesitant. Walking through a field immediately after a fire is something that most people from the western world know not to do: The smoke, the ashes and the heat are all still hazardous, threatening to seize your breath, make you dizzy and even lose consciousness. What was he up to?

Noticing our hesitation, he patiently explained, "The wind is blowing from us toward the elephants. If we walk through the fire, the elephants won't be able to smell us coming. If we take the path, they will catch our scent and most chances are that we will aggravate them. Believe me, you prefer the fire..."

Without further ado, we put our European ignorance in the hands of his African wisdom. Advancing as a close pack through the hissing, hot field was an unforgettable experience, overridden only by the close encounter with the elephants themselves (Photos 77–79, pp. 201–202). Using the literal smoke screen of the fire, we managed to reach the next safe camouflage offered by the trees without catching their attention.

While we felt euphoric as if we had just done something genius, Kendru only looked at us with a small, almost unnoticeable smile at the corners of his mouth. What was a completely new way of life-saving decision making for us was trivial, everyday intelligence for him.

***

To this day, intelligence tests are still used in a variety of contexts. While the relevance and especially legitimacy of IQ tests are debated, they are still considered a valid way to measure some forms of intelligence, particularly skills necessary for academic studies. We are still searching for ways to assess our diverse expressions of knowledge and problem solving, and how these can be taught.

Our complex display of intelligent behavior can be described as our tendency to flexibly develop and express our cognitive, emotional, creative and social functioning. It is as complex as the mathematical analysis of the number of combinations between multiple variables, displaying multiple values and influencing each other on the way. Subject to the culture we are born into, the society we live in and the genome we inherit, the possible combinations grow even further and the need to be able to use them in a multitude of ways becomes crucial. This head-spinning complex system of phenomena and action is the basis of human intelligence.

When we use our systems of intelligent abilities in combination with the human-specific system of the big five, we are expressing a kind of intelligence unique to homo sapiens alone: ***Technological intelligence.***

The emergence of technological intelligence on the stage of human evolution created an instrument of extreme biological advantage, resulting in the development of our continuously dynamic, persistent, ever-interacting nature.

In short, once we began putting our technological intelligence into practice, it wasn't long before we started changing the world.

Chapter 4

Enter, The King*: Technological Intelligence

"Computers," said one.
"Smartphones," said the other.
"Robots!" exclaimed another,
"A nightmare," sighed the fourth.

Every year, when I ask my students to say the first word that comes to mind when I say "Technology", the most common answers I receive are a list of various artifacts. The list usually includes the most relevant, latest bon ton in the sphere of high-tech gadgets. Some more traditional answers will include a variety of machines, equipment and gear. Even the most philosophical among them, who go as far as saying "the man-made world", usually seem puzzled when I ask them if they consider a book, a chair or a pencil as part of technology. Most of them don't.

What is technology? When did it begin? What kind of a relationship do we have with technology? How is it connected to the big five?

---

* **Enter, The King** was a ritual of introducing and paying homage to the King's royal entry during medieval times. Today, it is used as an idiom to introduce a dramatic presence/entity.

Do other organisms "have technology"? Why are these questions important in our journey to the uniqueness of homo sapiens?

To begin our discussion, let's explore the context of technology through some examples. We'll begin with the biological ability of living organisms to not only move but to actually know where they're going: The ability to navigate.

I Like to Move It[1]

Movement from place to place, even if the scale of space and time is extremely minute, is one of the characteristics of life. Organisms move, changing their spatial location using a variety of processes. A result of the interaction between physical features and habitat conditions, the variety of movement types is remarkable. It ranges from microscopic amoeboid movement or flagella-driven locomotion of microorganisms to tropism,[2] sliding, contracting, swimming, walking, hoping, climbing and running. The living world is constantly on the move.

One of the most entertaining movements I have witnessed is the Dance of the Sifaka Lemur (Photos 80–86, pp. 203–206). Enjoying the unique wildlife of the Berenty Reserve in southeast Madagascar, I noticed a small party of beautiful, mostly white, strong-legged Sifaka Lemurs. They were carefully situated on the trees, looking at each other and at the open terrain they were about to cross. On the other side, high trees bearing tasty fruit and bark were the table they had ordered for lunch. After a few moments of thought and exchange of looks, the die was cast. Navigating his way through the high grass, the first one started leaping and dancing toward the trees, and the others followed his mark. As if guided by music and a compass, their dance was on beat, graceful and exactly on the mark, making my own movement look awkward and clumsy.

---

[1] **I like to move it** is a 1994 hit song by *Reel2Real and the Mad Stuntman,* Erick Morillo and Mark Quashie.

[2] **Tropism** — the movement of plants as a result of stimuli such as light or gravitation.

Watching Salmon fish jump their way upstream was no less fascinating. Salmons navigate their way back from the ocean, sometimes through thousands of miles, returning to their home stream to spawn. Standing on the bank of a stream in Alaska, I was mesmerized by the beauty and strength of the "Salmon rush", the local name for this phenomenon of nature. A flow of Salmon made their way upstream, struggling against the current, determined to find their way using the earth's magnetic field combined with smell recognition of their home stream. Exquisite.

Organisms have an impressive variety of ways to navigate their movements, including the use of landmarks, ocean currents, the stellar skies, communication methods and magnetic fields, usually using a combination of several methods. Examples are wide and rich. There are indications that some birds can migrate throughout the globe due to a biological "GPS", magnetite crystals that are placed in their beaks. Mallard ducks can find the north using the stars. Grey whales navigate according to the topography of the ocean floor and wildebeests are thought to follow the scent of rain and chemistry of grass.

Certain fruit bats are known to travel thousands of miles in the dark, arriving at a very specific tree they are searching for. Every evening, fruit bats that live in a cave about 12 miles away come to feast on a fig tree right next to our offices. While in our eyes this fig tree is no different from thousands of others found much nearer their cave, these bats insist on ours.

Monarch butterflies display impressive cross-generational navigation. The butterfly leaving Canada will be the great-grandparent or even great-great-grandparent of the one who will finally arrive in Mexico, yet every year they arrive precisely at the same place that their ancestors left.

Specific navigational methods differ, but one thing is clear: The biological navigation capabilities of most organisms are much more developed than those of humans.

What we do have is our biological ability to imagine, explore, develop and put to use sophisticated *external* means, also known as technological artifacts. We use these abilities to not only compensate for our poor biological navigation systems but to expand and change them entirely.

If you don't know where you want to go, it doesn't matter which road you take, the Cheshire cat said to Alice.[3] He was right. For centuries, the ability to accurately plan our movement, knowing exactly where we are, where we're headed and how to get there, was one of the most challenging issues we faced. It directly influenced every aspect of life, from settlement to trade, social relations, politics, conflict and so on.

On land, known and stable landmarks such as hills and rivers help us map and find the way. But, in featureless landscape, such as vast deserts or the open seas, other frames of reference are necessary.

The solution was first developed as far back as the 3rd century BC in ancient Greece. Locations on earth are identified through a systemic, artificial grid of lines running east to west, called latitudes, and north to south, called longitudes. Independent of earthly reference, it is a beautiful, intelligent technical system to this day.

Latitudes are designed to begin at the equator and run parallel to each other from equator to each of the poles, shrinking on the way. They can be measured in accordance with natural movements and phenomena, such as the length of daylight, the height of the sun or the display and changes in the stellar sky. They are therefore considered "nature's decision" harnessed by man. Consequently, the definition of the equatorial latitude as the beginning point of latitudes is a "natural" point of beginning, and latitudes are easily found and accounted for.

Longitudes are a very different story. They are same-length circles stretching from pole to pole and back, created by man's imaginative drawing alone (See Figure 1). As a result, longitudinal locations have no natural point of reference, on earth or sky, that can assist in measuring them along the way. Finding an accurate way

---

[3]From the 1865 novel by English author Lewis Carroll, **Alice's adventures in Wonderland**.

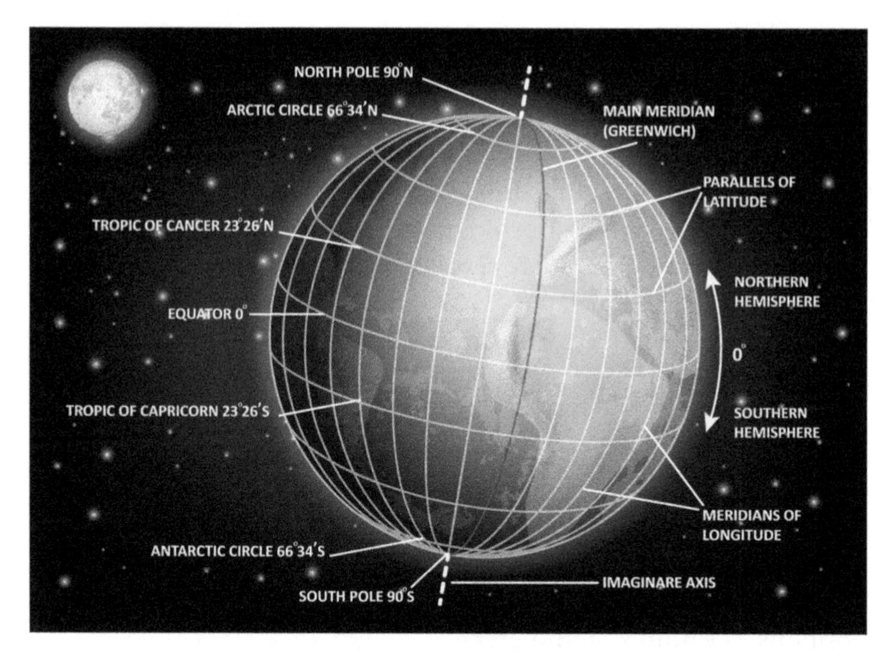

Fig. 1. Longitudes.

to determine longitude relied solely on the ability to accurately measure time.[4]

Using the grid on land was easy enough. At sea, however, it was extremely tricky, but crucial for calculating exact speed and position. If you can't accurately measure your longitude, you will easily lose your way. This frequently happened to many a good sailor, resulting in loss of lives, as well as merchandise, goods and ships.

It took centuries of study, trial and many errors to solve the longitude problem. Some of the greatest scientific minds were involved in trying to meet the challenge, all in vain, mainly because they were "looking under the streetlight" instead of searching for

[4]The full story of solving **the longitude problem** is told in detail in numerous books — one mentioned in the "Further Reading" section at the end — as well as videos, articles and movies. Here, a brief description of it is mentioned as one brilliant example of technological intelligence. I strongly recommend learning its fascinating and spirit-lifting details.

an entirely new perspective. The answer finally came from a self-educated, creative, stubborn and brilliant technologically intelligent Yorkshire carpenter and clockmaker named John Harrison. He spent his entire adulthood building prototypes of chronometers, perfecting them one after the other through multidisciplinary use of chemistry, mechanics, horology, metallurgy and the kind of commitment necessary for such revolutionary achievements. H-4, his fourth chronometer dated 1759 and resembling an oversized pocket watch, was to become the masterpiece that solved the longitude problem.

Since H-4, we have developed other methods and devices to navigate the world. Today, latitude and longitude are both determined electronically through advanced forms of GPS — a worldwide system of navigation including constellations of satellites and ground stations. With these artificial objects that we have placed in the sky as "stars" of reference, we are able to calculate terrestrial positions to within a centimeter.

When I heard that Harrison's chronometers — or as he called them, "time keepers" — were on display at the Royal Observatory in Greenwich, there was no doubt in my mind: I was determined to see them when the first opportunity would come my way.

Standing in front of these brilliant products of human triumph over its own shortcomings, I was amazed by the fact that they are still working, overwhelmed by the precise, delicate excellence of these works of technical art.

Watching their gentle, rhythmic movements, I kept going back to Harrison's own words upon the completion of H-4:

> *I think I may make bold to say that there is neither any other Mechanism or Mathematical thing in the World that is more beautiful or curious in texture than this my [longitude] Time-keeper.*

He was right. His chronometers are a source of inspiration that travels far beyond themselves, much more than the actual, specific artifact. They are one of many examples that are the result of the elegant merger of human intelligence, craftsmanship and artistic beauty. In my opinion, the best way to describe them is as a beautiful example of human ingenuity, an exquisite manifestation of what I will define as our *technological intelligence.*

Evolutionarily developed and intertwined with our big five, technological intelligence is the unique, specific system of capabilities that enables us to develop means to design, alter and radically change every aspect of our lives. In this case, it served to enable us faster and more accurate movement, on land, sea, air and space, taking us off our feet and into trains, cars, ships, submarines, airplanes and, more recently, the space shuttle.

The same realization came to my mind when I was standing on the edge of the Hoover Dam, exploring Captain Doug's radar or appreciating WEIZAC, the first computer in Israel, and one of the first large-scale electronic computers in the world. Witnessing the representations of our best, most important capabilities creates a rush of admiration and pride. Our technological intelligence is what has taken us from hunting to refrigerators; from using the light of a campfire to building a dam that brings light and power to more than a quarter of the populations of California, Nevada and Arizona; from naked eyesight to increasingly powerful magnifying microscopes and telescopes, enabling us to reveal and conquer the tiniest as well as most remote exciting, new worlds.

Technological intelligence. It's the kind of realization that makes a heart soar.

Our technological intelligence is one of the most important features developed by evolution in homo sapiens. It is one of the characteristics that define us as humans, the very game changer that has enabled us to take our story to places that no other organism has reached. It is innate and an inherent. An ongoing, biological combination that works in synergy and continuously drives us to break the boundaries of our biological capabilities, extending them far beyond our physiques.

We use it to make admirable achievements.

It's also what enables us to bestow unprecedented, sometimes irreversible, harm.

It places in our hands both incomparable power and extreme responsibility.

It's a definition that changes the very way we view a central component of our lives, one that we feel we are familiar with: *Our technologies.*

When I Created a Hammer[5]

What is "technology"? Do other organisms "have technology"? Why am I insisting on the term "technological intelligence" and what does it mean?

Considering the centrality of our technological abilities in our lives and history, the diverse paradigms through which we perceive them is somewhat surprising.

Probably the most widespread paradigm claims that technology is sophisticated tool-making. Some place the focal point on the process; others will focus on the product, whatever tool, machine or artifact that product may be. In both cases, tools are defined as objects that we use, modify or create to extend our biological abilities, changing the interactions between ourselves, other organisms and the environment. The focal point of this definition is external, elevating technology and what it enables us to do by giving it an independent identity and purpose.

This approach is easily found in the semantics we use when we verbally discuss or write about technology. "You won't believe what this computer did to me!" we exclaim when something is wrong on our screen, as if it was an entity with a mind of its own that is set out against us, and not an ingenious system that we ourselves built and designed to operate according to whatever we decided to put into it. "And then, rafts appeared" is a sentence I found in a schoolbook describing naval history, as if rafts can "appear" on their own and not as a man-made creation. Some even go as far as discussing technology as something that co-evolves *alongside* human beings and has as much as political agendas of its own, while others claim that our big inventions, such as the printing press, nuclear energy, the

[5] Paraphrase of the 1949 song **If I had a hammer**, written by Pete Seeger and Lee Hays, in support of the civil rights movement.

Internet and so on, have upsides as well as downsides, as they make new things possible while creating chaos.

The truth is that our inventions don't have any kind of side. *We* do. We are the ones who decide how to behave, including the decisions regarding when, why and how to use the technologies we develop, and which agenda — political or not — to promote through the use of technologies. Our decisions are based on our culture, beliefs, values, norms and goals. One group's possibilities will be another group's chaos. The most influential global prize we have, the Nobel Prize, is a result of exactly that.

Viewing technology as having a presence of its own leads to the popular arguments claiming that many animals "use technology", as if it were water or earth that can be used.

Agreed, other organisms show preliminary use of artifacts as tools to solve problems and communicate knowledge and behavior traits to their peers and offspring. For example, some dolphins use sponges as protective objects, "wearing" them on their beaks like gloves to protect themselves from injury by rocks, coral or other harms while scavenging for food at the bottom of the sea. Beavers are thought to be second only to humans in their ability to manipulate objects and change the environment. Great apes are known to fashion sticks into "spears" and hunt small primates with them, and so on.

However, while other organisms modify objects to achieve goals, they lack three major capabilities that would render their tools the definition of technology, let alone their tool use the process of technological intelligence:

- The ability to **produce complex modification or creation** of entirely new objects;
- The ability to **continuously develop and improve** their tools in an ongoing fashion; and
- The ability to **transfer** the complex capabilities they achieved **from one generation to the next.**

Lacking these processes, rapid change of circumstances or of environmental factors might leave these organisms endangered, even extinct. In contrast, homo sapiens have mastered them, as we will soon see in every example we will explore. So, while we may not be the only tool users, we are definitely the most sophisticated, advanced player in the field.

Another issue with viewing technology as having a presence of its own is that it easily leads to the kind of adoration and deification that we have witnessed during the 20th century, when it was widely believed that all human problems would be solved by technology. This approach inevitably led to disappointment, disenchantment and even demonization, as every technological development led to the creation of new problems and needs.

This isn't a surprise anymore. Suffice it to look at the growth of drug-resistant bacteria, a direct result of our development and use of antibiotics, or at aquifer pollution, a result of extended fertilizer use, or at the fact that before we were even able to finish saying the sentence "COVID-19 vaccination", there was news of new mutants that might need a different solution to cope with.

More than this spiral of problems, the issue that we have yet to solve is how we make sure that our technological solutions are also sustainable ones, taking responsibility not only for our human perspective but also for environmental perspectives.

That, in a nutshell, is the most important, pressing question of our time. By regarding technology as an entity on its own, it is easier to place the responsibility for the consequences of the use of technologies on the products than to take responsibility for what we do with them. Claiming that contraceptive pills and devices are responsible for reckless sexual behavior or that social networks and discrete relationship sites are the reason for increased divorce is misguided at best. To say that weapons are responsible for violence or that the accessibility of violent computer games allegedly breaks down a normal sense of conscience in younger people is no more accurate than blaming the messenger for the message. They are all examples of the result of values, norms and decisions that are the prism through which we use these technologies and control the way others use them.

In other words, plastic isn't responsible for polluting the environment, *we* are.

Another definition, found in a multitude of dictionaries and textbooks, claims that technology is the result of science application. Even a distinguished, highly valued source like the Encyclopedia Britannica defines technology as "the application of scientific knowledge to the practical aims of human life or, as it is sometimes phrased, to the change and manipulation of the human environment", suggesting that science inquiry and the resulting scientific knowledge are the driving force underlying the creation of machines, equipment and practical products of all sorts.

If we consider science as the empirical, testable, evidence-based school of methodologies and knowledge that have developed mostly since the 17th century, then viewing technology as "the application of science" couldn't be farther from the truth. Technology has been a crucial, inseparable part of the development of mankind long before science was developed. A brief overview of the history of technological developments, such as means of communication, travel and healthcare, reveals that they were successfully developed and used long before they were explained by the scientific knowledge necessary for their full understanding and further development.

Think, for example, about the expansion of our ability to use one of our basic biological senses: Our ability to see.

> Letters, however small and indistinct, are seen enlarged and more clearly through a globe or glass filled with water.

This observation, made by Roman philosopher Seneca in the 1st century AD, is how eyeglasses first started. It had nothing to do with science. Without any knowledge of the intricate workings of the eye, the brain, the properties of light, glass and water or the interactions between them, Seneca's testimony had to do with a simple, everyday observation and is the first recorded applause to magnified sight.

It was a whole millennium before the Arab scholar and astronomer Ibn al-Haytham suggested that smoothed glass might assist someone suffering from visual impairment, and another 200 years before his idea was implemented for the first time, when Italian monks developed the first "reading stone": A semi-spherical lens made of rock crystal and quartz that, when placed on a piece of writing, magnified the text. Significantly improving the monks' ability to read, consequently improving their quality of life, "reading stones" were the first mark on the road to the development of eyeglasses.

The invention of "reading stones" followed by eyeglasses in the 13th century wasn't considered as an "application of science", but as a form of technique art, a term used then for technology, made for achieving good vision.

It should be noted that there is evidence of scientific knowledge incorporated within technological development in ancient times. Admittedly, Sumerian astronomers in Mesopotamia and Mayan astronomers in Central America developed calendars, time-measuring devices and irrigation systems based on observing and plotting the motion of the celestial bodies with remarkable accuracy. However, "applying scientific knowledge" wasn't the driving force of their technologies. Their stories were.

To this day, the archeological remnants of the temples at the national park of Tikal (Photos 87–89, pp. 207–208) in Guatemala are an impressive as well as gruesome reminder of the fact that the use of our technological intelligence is subject, first and foremost, to our belief systems, values, norms and culture — in short, the stories we live by.

The name "Tikal" means "the place of the voices". Visiting the temples raises the feeling of how appropriate and gloomy it can be. It has double meaning when standing at the top of one of these high, majestic temples, above the tree canopies. The heads of the other temples stretch through the greenery to the sky. The voices of the howling monkeys crying from the trees sound like imaginative, haunting voices of young Mayan women and men from past centuries, who are believed to have been sacrificed here to the gods.

The technological abilities of the Mayans were remarkable, primarily based on astronomy and mathematics. Seven temples in one of the squares were designed as highly sophisticated astronomical instruments, used to measure and count days, weeks and even the months. This was necessary for planning the year, agricultural cycle and traditional festivities, including special festivities that took place once every 20 years, marking the Mayan story of creation.

What exactly happened during those festivals, when and why human sacrifice was made, is still under research. Evidence such as art, writings, ornaments and body remnants have led to theories claiming that those who lost the special 20-year festive ballgame were sacrificed.

The complex Mayan system used two calendars. Once every 52 years, these two calendars coincided, an occasion that the Mayans believed opened the gates to heaven. This event set the stage to sacrifice the winners instead of the losers, as the celestial event had won them the privilege of immediately entering the next world.

In spite of their advanced, deep knowledge of astronomy, the Mayans still believed that human sacrifice was necessary to please the gods because without it, the sun will cease to rise.

The scientific knowledge the Mayans had wasn't the driving force of their technology. Their belief systems and culture were.

Toward the 17th century, the interaction between technological innovation, scientific understanding and their joined impact on society attracted the interest of philosophers, scientists and craftsmen, who began to view, discuss and develop means to enhance our mastery of nature and the influence of this mastery on our lives. This created an amplified merger of science and technology in connection with human needs, a process that since the 19th century accelerated at an unprecedented pace.

It is a union of reciprocity:

- The development of complex technological systems is dependent on scientific understandings and methodologies.
- Complex scientific research depends on the use of sophisticated technological systems to advance.
- Society continuously changes under their realm.

Science, technology and society are no longer separable.

Now that we have explored our relationship with technology, let's dive deeper into the development that enabled it.

Technological Intelligence

The development of technological intelligence in humans marks the crossroads at which we took a different path from the path taken by the rest of life on earth, thus becoming the driving force of our evolution.

Looking at it through the prism of the big five gives it full meaning. We consciously act within the kaleidoscope of imagination combined with creation, continuous development over time, extending the boundaries of physique, consciousness, emotions and behaviors, and all this, while symbolizing and communicating these achievements to others of our species in ways that break the boundaries of time and generations.

It is through this spirally evolving process that tool use became technological intelligence, the engine we use to change the world.

Of all the living organisms on this planet, only homo sapiens possess it.

Considering that we share some 99% of our DNA with chimpanzees, the impact of that mere 1% difference is mind-boggling. If there are similar living beings elsewhere in the universe, who are only 1% more developed than us, it should be no wonder they can easily hide from us. We are no more than a chimp to them.

Hopefully, if we do meet them, they will regard us as a cute chimp.

Back to ourselves. It is hard, maybe impossible, to point the exact time and process through which technological intelligence developed, but it would be an educated assumption to begin looking at it when our ability to control natural forces, such as fire, began to develop. Accumulated fossil, genetic and archeological findings suggest that hominids began to control fire as far back as 1.8 million years ago. The control of fire is connected to the beginning of food modification by cooking and, simultaneously, to the increase in brain size, a decrease in digestive track and the changes in jaw and teeth sizes which occurred around this time. It results in the fact that other primates' diets are insufficient for us, as we can't extract from them enough calories to live healthy and active lives.

Cooking isn't just about combined changes in physiology and diet. It also involves the development of tools, modification, storage and preservation systems. It requires social and personal interactions, such as communication, learning and trade. Through a cumulative process of trial and error, reconstruction and recombination over a period of only several hundreds of thousands of years, we have gone from scorching meat on fire in the savannah to taking lunchboxes to the moon.

To develop our technologies, we combine innovative creativity, basic and high order thinking skills, ongoing logic, design and doing with the use of natural materials, phenomena and effects. These are all part of the complex systems of intelligent behavior, in this case — leading to technological abilities.

Let's return to the example of our eyesight. We have come a very long way since the "reading stones" of Italy. Our growing understanding of light and of the optical properties of materials, combined with accumulating knowledge about how our eyesight works, the manufacturing and modification of plastic and metal, the ergonomics necessary for comfortable eyewear and so on, led to continuous improvement of eyewear solutions.

Now, let's think about the change in our eyesight as we age, at which point most of us need multi-focal eyeglasses. Through flexible administration of additional knowledge and know-how, we have further developed eyewear that continues to expand the biological ability of our aging eyes; and through medical developments, there are many cases where we are able to alter the structure and function of the eyes themselves, avoiding the need for eyewear altogether.

Take this a few steps further, and think about the expansion of our optical abilities not only by eyeglasses but by devices that expand our boundaries even more: Telescopes and microscopes.

The first use of telescopes was by pointing them at the horizon. Then, through combining creative thinking and innovation, in 1609, Galileo turned his telescope toward the sky — and a whole new, distant world was discovered, opening a whole new frontier to be explored.

Similarly, during the second half of the 17th century, curiosity and high technical expertise made Leewenhoek turn his microscope on a drop of water — and a whole new, tiny world was discovered and opened yet another frontier.

Both frontiers, the extremely far and extremely small, have become resources of innovation and expansion, such as satellites, space shuttles, antibiotics, vaccines and so on.

These examples and more are all results of our intelligent creating abilities. They were designed to answer our curiosity and craving for ongoing change, enhancing our ability to take the next exciting step in the spiral of human generations. Year after year, decade after decade and century after century, we rely on our technological intelligence, driven by motivation, curiosity and persistence, to continuously develop and create new combinations of innovation in new and flexible ways.

We will likely never cease to use it to develop new technologies. It's who we are.

Homo Sapiens' creativity, problem solving and innovation are truly a beautiful tapestry of growing complexity. Albeit a purposeful process, it is interesting to observe the way we develop our technological inventions through the lens of the major forces that drive evolutionary processes: Diversity, selection and ongoing development.

Here's an example: We continuously develop different means of personal communication. Cellphones of all sorts, mini-phones, tablets, mini-tablets, smartwatches, you name it. We select the specific gadget we need to meet our needs, from chatting to texting, twitting or tic-toking — Usually more than one, for all of the above. Every year or so, we eagerly wait to hear announcements of the new-and-improved versions, rushing to check and purchase them, most of the time for minute, not to say negligible, reasons. Sometimes, what we are using today is very, very different from what we used merely a year ago. That's how ongoing development of communication works.

Another example: When I was 6 years old, my father took me to choose my first camera. I remember choosing from a wide variety of different cameras as each had slightly different features than the other. I finally selected the one that had the most important trait: It was pink and covered with figures of Minnie Mouse. It required film that, after shooting pictures, had to be taken to a photography shop where I left it for a few days to be developed into hardcopy pictures. Those days of waiting were spent hoping that most pictures would come out as I had hoped they would. I would then have to store the developed film, keeping it out of harm's way so I could use it again to make duplications of the photos I desired.

As years passed, the criteria I used to select the next camera changed and developed together with my own growth, more sophisticated selection opportunities, changing needs and goals. The time and process consolidating changes first made to cameras eventually led to the developments and changes made to computers, videos, telephones and a combination of them all. After many years and marvelous innovations with increasing ease and usefulness, we arrive to the present era, offering experienced photographers as well as novices of all ages a much wider range of digital operations and achievements. Wait a few years into the future, and you are sure to experience the next steps in the evolution of our photographic and video abilities.

Using the lens of evolutionary criteria on technology leads to interesting observations. One more example: Our craving for diversity has propelled us to create an array of shoes. I challenge you to prove that we really need so many different shapes, colors and kinds of shoe designs, styles, heel heights, sandals and boots. The ever-increasing amount is in part due to the selection of the fittest for a specific purpose. Agreed, running in high heels will not win you a race, is not a great idea in general and definitely not recommended when you need to escape danger. Don't forget ongoing development of fashion, styles, new needs and so on. We can join the choir: We always, but always, need a new and improved pair.

Every once in a while, a revolutionary technology will be developed, changing the evolutionary course of an entire area of human endeavor, as was the case of Harrison's chronometers and many other inventions throughout human history. Such were the printing press, the incandescent light bulb, the World Wide Web and the example of **the horse manure crisis**.

In 1898, the world's first international urban planning conference in New York held a debate about a pressing crisis that was threatening large cities all over the world: What shall we do with — excuse my language, but this was the real problem — accumulating horseshit. Literally.

Toward the turn of the 20th century, cities around the world were swarming with cabs, carts, buses — all driven by horses and depending on them for the transport of people and merchandise. Horses by tens of thousands per city, per day, were an issue of major concern. Manure and urine were one part of the problem, creating major sanitary and health problems spread by flies and disease-inducing microorganisms. Horse carcasses were another, often left to decay before removal, leading to yet more sanitary and health issues. A potentially poisonous situation, it was a growing problem in every city, from New York to London to Amsterdam. In 1894, it was declared by the *New York Times* to be "the Horse Manure Crisis", leading to the 1898 international 10-day conference that was set up to try and find solutions. After only 3 days of debate, attendees came up with no solution, leaving the conference frustrated and worried.

The answer came within less than two decades from an unexpected area, rising from the human driving force to continue to imagine and create new ideas and build new devices. In this case, it was the invention of the assembly-line, fit for every pocket, mass-manufactured motorcar. By 1912, cities around the globe replaced horses with motorized vehicles that became the main source of transport for people and goods alike. Goodbye horses, hello fossil fuels.

Which isn't a final point either, as the nature of our species demands. As the environmental and severe health problems caused by contamination of air, soil, water, food and body systems by the use of fossil fuels increase, we continue to search for ways to create more environmentally friendly cars, such as through the use of unleaded fuels and the development of electric or hybrid vehicles. The hereditary fashion that we utilize when we develop our technologies ensures that as we continue to design the next generation of transportation; it will rely on and improve the technology we already have, without going back to horse and buggy all over again.

It is Automation, I Know[6]

Observing our story from the beginning of the agricultural revolution, some 10,000 years ago, to date, proves that our technological intelligence took the central role in crafting our development as societies and cultures. It has enabled us to radically design and change our relationship with the most important realms of our existence: Time, space, communication, health and knowledge. Our continuous need for diversity, selection and development of "the next best model" that will expand or replace our biological abilities in every realm is insatiable. So much so that I am often asked whether I think that the technologies we develop will someday be so sophisticated that they will overcome us entirely. Can our technologies arrive at a point where self-design, self-improvement and self-reproduction take control? Will they be able to throw us aside, enslave us or, worse than that, get rid of us altogether?

The idea of possible independence of our own creations isn't new, as can be found in ancient and medieval stories, such as Ovid's Pygmalion[7] or the Golem of Prague.[8] It is a reasonable idea if one views technology as a stand-alone identity, with a mind and

[6]From the 1963 song **Automation**, by American song parodist **Allan Sherman**, dealing with the threat people feel when encountering advancing technology that might replace them.

[7]**Pygmalion** — in his book *Metamorphoses,* roman poet Ovid tells the story of Pygmalion, a sculptor who creates an ivory statue of the perfect woman only to fall in love with her. Aphrodite, the goddess of love and desire, answers his prayer and brings the statue to life.

[8]**The Golem of Prague** — the 16th-century tales of the Rabbi of Prague, who created a figure made of clay and brought it to life to defend the Jewish congregation, eventually losing control over it. The independent Golem became violent, even murderous, before the Rabbi was finally able to immobilize it, after which it disintegrated.

responsibilities of its own. The path from such a perception to demonization is as quick as the path from the same perception to deification. The term "Artificial Intelligence" ascribes exactly that: Autonomy of our advanced technologies. Along with it come both the element of fascinated, optimistic imagination waiting for a new and improved world and total despair stemming from pessimistic exaggerations of the power of technology to determine its own course apart from any form of human control.

Defining technology as the systematic school of knowledge for making and doing, resulting from our technological intelligence, is a different conceptual framework. It places technology as a human-dependent phenomenon. As such, it cannot possess complete autonomy. Technology cannot be perceived as an identity on its own, for one basic reason: Its very existence has one distinct requirement — the ongoing existence of a species capable of sustaining it.

Simply put, without us, it won't work.

A pressing issue lay elsewhere: The growing gap between our rapidly developing technological abilities and the relatively slow-paced development of our socio-emotional ones.

The Tinman's Heart[9]

The observers were curious. A scientist stood on stage, next to him a woman was seated. Her clothes suggested she was a nurse.

The scientist asked for a volunteer, and one of my colleagues stepped forward.

"What is your name?" the woman in the chair asked him.

"Dan," he answered.

"Hello, Dan. Where are you from?" she continued.

"Florida," he replied.

"That's far!" she exclaimed. "Thank you for coming to visit us here in Japan. Can I offer you some refreshments?"

---

[9] **The Tinman** is a character in L. Frank Baum's 1900 book: *The Wonderful Wizard of Oz.* In the book, *Dorothy,* the protagonist, is on her way to meet the Wizard in the hope he will help her return home. The Tinman joins her on her journey, so he can ask the Wizard to give him a heart.

The conversation went on for a few minutes. We were at a conference displaying new technologies that use artificial intelligence, and the nurse in the chair was a black-haired, Japanese-looking robot.

The scientist on stage was very proud. He described in detail how this robot will help us learn about the advantages of robotic healthcare, and more importantly, how artificial intelligence can replace us in many of our chores. The nurse-robot was already successfully used in homes for the elderly, as caregivers and companions.

When he finished his presentation, the scientist asked if there were any questions.

A woman — a real one — from the audience raised her hand. She sounded distressed.

"Where does her voice come from?" she asked.

Puzzled, the scientist tried to answer. "There's a microphone in her throat..."

"That's not what I mean," the real woman persisted. "That's technical, and I get it, but..." she paused for a second, in search of words. "Her voice, not the *technical* kind, the *human* kind...the spirit, the care...the voice that comes from the heart, where does hers come from?"

The scientist dismissed her impatiently. "I'm sorry, but your question has no meaning and is therefore irrelevant. Anyone else?"

I understood where her anxiety came from. It had nothing to do with technology. It had to do with *people.*

Let me explain.

The 20th and 21st centuries are identified, almost immediately, with the term "technology". The technology we have developed throughout them has changed at a pace that has never been seen before. Within less than 200 years, more than any equivalent time frame in history, we have developed our capabilities via our technological intelligence in every single aspect of life. The scale and scope of change have accelerated to the point where innovation is no longer an occurrence, but a value in and of itself.

The examples are endless. We continuously design our environments and even our own bodies, changing our quality of life, its content and its longevity. We have established ways to make our fast-growing capabilities into available, free-for-all and sophisticated devices, created to automatically do many of our chores. Cooking, cleaning, heating, safeguarding and, as the example above shows, healthcare are easier than ever before, making our lives more comfortable and freeing our time to be used for more desired and enjoyable tasks.

We are more mobile and connected, with growing global and real-time dissemination of goods, information and interactions, found even in the most remote, unexpected areas (Photos 90 and 91, pp. 208–209). We can now react immediately to the needs of a community that was devastated by an earthquake, hurricane or tsunami on the other side of the world, sending lifesaving technical, medical and social teams to the rescue within hours. We can also react and demand the immediate stop of atrocities initiated and practiced in other places in the world, such as deforestation, fire-controlled agriculture, development of weapons of mass destruction and cyberattacks.

We are more productive and healthy, and are able to take care of our health and well-being better and faster than in any other time in the history of human development. A mere 20 years after my father passed away from heart failure, my brother-in-law was saved from the same fate by a mechanical heart implant; and when faced with the global COVID-19 pandemic, it took less than a year to develop vaccines that otherwise would require a process of several years. Our longevity, as well as quality of life, has meaningfully increased. "The child who will live to be 200 has already been born," claim several futurists.

The problem is that we haven't really stopped to consider what we will actually *do* with 200 years. What will life look like? What kind of jobs will we have when we live so many years? How will we prepare for them? How will we ensure sufficient income that is necessary for reasonable living throughout so many years? What will we do with the expanded leisure time we will have?

What about marriage? Is the spouse I married when I was, say, around 30 years old, going to be the same person I will be with for the next one hundred and 70 years? What will happen to family life as we know it? To friendships?

These are but a sample of questions we still need to address. We are becoming more aware of personal, social, economic and environmental outcomes and needs. They are all part of an additional, crucial aspect that the past century has painfully placed at the front: The need for ***sustainable development*** — The ability to meet our present needs without compromising the ability of future generations to meet theirs.

S is for Sustainability

Sailing swiftly into the King Edward Cove of South Georgia Islands toward Grytviken, the majestic scenery was breathtaking. The stunning bay lay at the foot of high snow-patched mountains, with stretches of greenery and rocks. The few buildings of the historic settlement, now inhabited only during summer, sat at the head of the bay, serenely waiting for us. The white church stood somewhat aloof, an obvious observer of the working parish that used to be nearby. The bay was peaceful, the water was calm and shining in the sun and our hearts soared by the splendor of it all. As we landed, king penguins, birds and seals dotted the landscape (Photos 92–96, pp. 209–211).

The beauty around us was about to be an extreme opposite to the cruelty that had unfolded nearby.

Walking toward the small harbor, evidence of past activity lay in the sun: A large ship named Petrel was beached, lying to rust on shore. I noticed large holes lined across her body. Next to her, large strong iron chains lay in high heaps (Photos 97–99, pp. 212–213).

In 1904, Grytviken was founded as a whaling station. Captured whales were used for different needs, or as whalers often bragged — "no part of the whale is unused". Supplying blubber, meat, and bones, the Petrel, like other whaling ships of the time, was at its full glory when it hauled into the station its full capacity of hunted whales. Since each whale was tied to one of the holes on the side of the ship by these heavy chains, I walked around the Petrel to count.

Fourteen Holes. Seven on the right, seven on the left.

Every time it sailed, the Petrel, ironically named after a free, sky-soaring seabird, was the death place of no less than fourteen captured whales.

According to local records, sailors from the beginning of the century attested that there was such an abundance of whales in the bay that during the first years there was no need for the ships to hunt in open ocean waters. The whales were right there, living their lives in the bay with no idea that they were about to be slaughtered. Thinking only of immediate profit, whalers hunted recklessly. No one stopped to think about the long-term ecologic, economic and ethic aspects of these actions. Sustainable thinking and compassion for the environment were not part of the whaling world.

Within 60 years, the whale population in the bay and even the ocean around the island reduced to such level that the bay fell silent, and whaling became useless. In 1966, the whaling station at Grytviken was shut down. During the second half of the 20th century, over-whaling and diminishing of whale species led to an almost global prohibition of whaling. Today, more than half a century later, the bay of Grytviken is still empty. The damage has been almost unrepairable.

Listening to the silence of the bay, I couldn't help but think about the human–animal war that took place here. Whales are intelligent mammals, with developed social lives, communication and cultural learning. It isn't surprising that they don't trust us anymore, staying as far as they can from anything resembling human beings.

The name we gave ourselves, homo sapiens — "The Wise Hominid" — didn't seem so right anymore.

Global risks arising from implementation of technological developments present challenges of long-term sustainable fashion. Climate change, clean energy resources, food security, water availability and health promotion are only part of these challenges.

Organizations, corporates and governments are searching for policy making, scientific and social solutions to handle these issues. This is particularly tricky since the beneficiaries of these solutions —

namely, our grandchildren — are not yet necessarily present, therefore the benefit for the investors — namely, us — isn't always apparent enough to motivate our actions. Although some countries established committees or other entities to address these issues, many times they were shut down when finances became problematic.

This situation brings to mind a beautiful story I once heard:

> "Come, child, and I will tell you a story you should always remember," said an Indian grandfather to his young grandson. The boy came close, sat next to his beloved grandfather and listened.
>
> "Two wolves live inside your stomach," said the grandfather, pointing at his grandson's belly. "One wolf," he continued with a happy voice, his eyes sparkling, "is a kind, generous and loving wolf, who creates new opportunities for you, helps you in your adventures on your way to achieve your dreams. It reminds you how to make sure you and your loved ones are happy and healthy, a good wolf who is satisfied when you have a prosperous, fulfilling life."
>
> The grandfather paused, his eyes darkened as he sighed and continued with a sad voice.
>
> "The second wolf is exactly the opposite. It is the wolf of cruelty, selfishness and hard feelings, complaints and violence. It is always using your dreams against you, inflicting pain and blindness toward others, harming you and your loved ones, mocking your weaknesses and satisfied only when you are miserable."
>
> "What are these two wolves doing in my stomach?" asked the boy in an uneasy, alarmed voice.
>
> "They are at constant battle," said the grandfather. "The two of them are part of every thought and decision you have to make. Through every step you take, they are continuously fighting each other, each trying to take control over what you do and who you are."
>
> Immersed in thought, the boy fell silent. His face looked as if he were concentrating on himself, trying to listen to the voices of battle going on in his stomach.
>
> After a few minutes, he turned to his grandfather and asked: "Which wolf wins?"

The grandfather smiled, and with a soft voice answered his grandson: "The wolf you feed."

***

Developing ways to use the power of our technological intelligence in a sustainable fashion is completely up to us.

Regulating the application of technology is crucial for creating a sustainable balance of natural forces, vital for our own existence. We know what needs to be done. We even know what kind of rules and behaviors we must administer.

There is only one problem: We're not very good at implementing them.

Like children, we are pointing at technology as a scapegoat for our shortcomings, blaming the tools we create instead of admitting that the most important value that our technological intelligence requires us to practice is **responsibility** — Understanding that the decisions about whether to develop something and use it, including the outcomes of this decision, are undeniably ours.

After everything else is considered, the one real question that we should remember to ask and learn how to answer is this: "Does the fact that I *can* do something mean that I *should* do it?"

There is only one way I know of that can lead us to meaningful and sustainable answers, based on knowledge, know-how and critical thinking. It's another human-specific endeavor that goes hand in hand with our evolution. It's also an area that, while being practiced for centuries and in practically every corner of the world, is still frustratingly, highly debated and misused. We will examine it in the next chapter: *Education.*

Chapter 5
The Fifth Freedom*

"This is not the way we do it."

Mrs. Tenenbaum's voice was trying to be tender, but her eyes were strict and determined. I could feel my face flush. As much as I tried to stand my full 8-year-old, third-grader's height, something in her stature made it clear that she was standing way over my head.

She was the teacher. She was in charge. I was no more than a disobedient bug.

A few days earlier, Mrs. Tenenbaum had given us an assignment in class: To write an essay about our thoughts regarding violence as a "means to an end". It was the end of 1968 and I was living in Los Angeles at the time. The war in Vietnam was at its peak. The trial of Sirhan Sirhan, Senator Bobby Kennedy's assassin, was discussed in almost every household. The topic of the essay was a frequent visitor at our dinner table, so I felt comfortable with the task. More than that, I was excited. Not about the subject, but about the fact that it was the first time anyone asked me to write about *my opinion.*

---

*In 1941, FDR's "state of the union" address defined the four basic freedoms each human being should enjoy. The speech has been since coined as "the four freedoms speech". After referring to them, the chapter will expand to discuss the fifth freedom not mentioned in the speech — education.

I approached the task with high motivation. At the end of my three-page argument, I wrote the final sentence — "…and everyone will suffer in the end". An enthusiastic reader, I remembered that at the end of all the books I read, after the story itself is done, the words "The End" appear — either as a separate sentence on the same last page or on a completely separate last page of their own. Those two words signaled that the story had come to a conclusion.

I looked at the sentence I wrote and decided to combine the two ideas in the same two words: The last sentence of the essay will also be the declaration of "The End" of the essay. Elated, I was convinced it would be a clever double use of the two-word text. I erased the two last words of the last sentence and rewrote them in the middle of the line below, like this:

…and everyone will be sorry in –
The End.

Mrs. Tenenbaum handed back the essay I wrote. There was a large red "B-" on it.

Being a straight-A student, I searched frantically to see what I had done wrong.

There was only one correction. Next to my artistic ending, Mrs. Tenenbaum wrote exactly what she had said: ***This is not the way we do it.*** Using the same red pen, she erased my ending, and rewrote the two miserable words at the end of the last sentence.

My heart sunk. What was the "it" she was talking about? Was there only one way "to do it"? Why? Who decided what that way was? Did she not understand what I was trying to do? Did she not see this small, non-conformist way to write "The End" as a creative expression of my own voice?

The look on her face left no room for discussion. I had obviously broken some important, unknown rule, and will have to be very careful not to do so again.

Since I was a good girl who followed the rules, it was a lesson well learned. At school, never again did I attempt to invent anything that wasn't exactly in line with what we were taught. I left my inventive, creative mind for other environments. It also showed me the influence that one sentence from one teacher can have on the life of a student.

Years later, when I became a teacher myself, I realized that Mrs. Tenenbaum had unintentionally taught me another important, not to say crucial, lesson: I would never say that sentence, or anything like it, to any of my students.

What's a Nice Girl Like You Doing in a Place Like This?[1]

Teaching and learning aren't unique to humans alone, as has been discussed in previous chapters. Acts of learning can be found at different levels of performance in primates[2] and other mammals, birds and even insects. Cheetahs actively teach their cubs to hunt, releasing in front of them live prey for them to catch, knowing that the prey might get away before the cubs learn to succeed. Cockatoo parrots have been reported to teach each other unique, newly developed foraging techniques. Ants display teaching abilities toward their peers when they actively lead them through newly discovered pathways to food sources, the leader making sure that the followers aren't lost on the way. Bees are impressive teachers, as they teach their peers exact directions to pollen and nectar by specific dances.

But, if you desire imaginative and continuous development, transferred from one generation to the next, teaching and learning are far from being enough. You need organized content, sophisticated means of communication, structures of knowledge and behavior transfer.

You need the big five.

---

[1] Paraphrase of **What's a nice kid like you doing in a place like this** — the 1966 song sung by Sammy Davis, Jr., in his role as the Cheshire Cat in Hanna-Barbera's version of "Alice in Wonderland".

[2] Noted, some of the most influential work that changed our perspective regarding the origins of human learning and teaching was that of **The Trimates** — the name anthropologist Louis Leakey gave to three women he chose to study primates in their natural environments: **Jane Goodall**, who studied chimpanzees; **Dian Fossey**, who studied gorillas; and **Birutė Galdikas**, who studied orangutans.

Homo sapiens are the only species that has taken the processes of teaching and learning much further than any other organism. We created strategic, educational processes through which we increase our ability to transmit, inspire and influence change in our intellectual, emotional, technological and social assets.

In short, other animals teach. Only Humans ***educate***.

Discussing the unique human endeavor called "education" in depth, with respect to the scope it deserves, will require an entire, separate book. In our case, I'll focus mainly on the highlights of its unique character in humans and how we should currently think about it by sharing some of my own experiences.

Some 22 years after my heartbreaking experience in Mrs. Tenenbaum's classroom, I was at the final point of my PhD studies in Molecular Biology at the Weizmann Institute of Science, Israel's worldly acclaimed leading science research institute. Realizing that my real passion was working with students, I handed in my thesis, received my coveted degree and, 10 years after I had left the high school classroom as a pupil, I returned — this time as a biology teacher.

In the 10 years that had passed, the world had changed. People began finding their way around using the help of GPS. The first space shuttle was launched. The first artificial heart was implanted in a human being. Microsoft filled the world with windows. The World Wide Web became a regular household inhabitant. We literally started making babies in test tubes by means of in vitro fertilization. We abandoned record players and embraced CD players, later abandoned for Bluetooth speakers and such. Everything was on the move and changing. Changing fast.

Almost everything. I walked back into the school classroom, and it was as if time had frozen and I had never left. It was exactly the same as I remembered it: A blackboard, chalk and eraser; a teachers' table in front, facing rows of student tables and chairs. Twenty-five students were facing me, eyeing me suspiciously. Some with eagerness, some with disrespect, most with indifference. It was just the same as my own high school. In fact, it had been like that for decades. Probably more.

During that very first week back at school in 1991, two things happened that became turning points that were about to change not only my career, but the very paradigm I held regarding education.

First, the principal of the school called me into his office for a meeting. Looking at my credentials, he said, "Impressive. I believe we can both agree that you are an expert in biology."

He then paused for a split second, just enough time to look me straight into the eye, and said, "Now, go study education. It's a profession as well."

The message was loud and clear: The fact that you know biology through-and-through isn't enough to make you a biology teacher. At least, not a *good* one.

Luckily, I listened. He was right. Education is one of the most complex, fascinating, ever-developing multidisciplinary professions, and probably the most disregarded one. Since then to this very day, some 30 years later, I am still studying the profession of education.

The other meaningful turning point came when I taught my very first 10th-grade biology class.

Naturally, I began by introducing myself. I watched as the 25 pairs of eyes in front of me grew wider. When I mentioned that I had just received a PhD degree, one of my students couldn't hold it in anymore and exclaimed, "What is someone like *you* doing *here*?"

In one short question, he revealed a sad, not to say tragic, observation: He had completely assimilated and taken for an unquestionable truth a message that the education system had been telling him since early childhood: If someone is considered a highly educated successful professional, he or she will seek for much more than becoming "only" a school teacher. Education just isn't really that inspiring, not to mention important, productive or financially rewarding, to actually be an appreciated and desirable area of expertise. Worse than that: "ordinary" school students like him are not worthy of well-educated teachers.

He was only fifteen, caught in a system that, in his eyes, failed to see who he was, what he needed and what school should really give him.

When we compare our education systems with the major characteristics of the 21st-century society, the same society that these systems are part of and are expected to prepare him and his peers to be part of, it's obvious why.

They Changed Society, But Didn't Tell Schools

The third decade of the third millennium is well on its way, and homo sapiens have never had so much influence on themselves, on other organisms and on the environment.

With the information revolution at unprecedented heights, access to best practices as well as to seemingly esoteric, minute or remote new knowledge is easily available to any inventor, engineer, scientist, artist, philosopher and educator.

Our databases, search engines and means of communication have outgrown the boundaries of geography and time. Our ability to generate and mold knowledge to fit our needs is easily available, fast and practically free for all.

In every corner of the world, people of all ages carry in the palms of their hands a small device called a cellphone or a smart phone. It has more information on it than any library could ever have, in multiple forms of media. Available in all languages at all times and almost all places, it also enables us to transmit these forms to whomever we desire, in real time. We can easily and immediately reach, communicate with and trade on a local and global level alike. As a result, we live in an era that is rapidly creating new tasks, occupations and possibilities that we can't always predict in advance.

Our leisure time has changed as well. Massive amounts of entertainment options are at our fingertips. Streaming devices are replacing movie theaters, concerts and cable television. Social Networks are core means of information dissemination, and communication. E-commerce is growing, leaving store shopping behind. Texting and Zooming are replacing interpersonal phone calls, physical meetings and even e-mails. We have learned to expect that the next new advancement is around the corner.

We also know much more about ourselves than ever before.

On the ***personal*** level, there is growing understanding and acceptance of the multitude of combinations that make each and every one of us unique. Personalities, tastes, tendencies, learning styles, competencies and needs work together and are expressed differently in each of us. More than ever before, we are coming to acknowledge that each of us has his and her "special needs".

On the ***cultural*** level, we are more familiar with the diversity between groups. We explore our own specific cultural contexts as well as those of other cultures, recognizing similarities and differences, swaying on the range from alienation to embrace and celebration.

On the ***universal*** level, we know more about who we are as a species, what our abilities, strengths as well as weaknesses are as one of the living organisms on this planet. As I have emphasized in previous chapters, we also know much more about how we interact with other organisms and the environment, for better and for worse.

Now, think about our schools.

Throughout the greater part of the world, be it Europe, Asia, the Americas or Africa, amongst the stunning variety of architecture, art, worship rituals, customs and social structures displayed by different cultures, one structure is steadfast and unchanging. Regardless of context, I can always easily recognize it: The structure that functions as a school.

Almost everywhere, be it a metropolis like Manhattan or a remote mountain village in Thailand, school structures have the same features (Photos 100–107, pp. 213–217). Classrooms look the same, their goal is mostly knowledge transfer and the means to do it are similar. Rows or aggregates of tables and chairs, a place at front for the teacher — almost always a woman — and a device considered as a crucial knowledge transfer vehicle, such as a black board and chalk or, in more "advanced" settings, a white board and a marker, or in even more "updated" settings, a screen and a keyboard.

Fig. 1. Eitan Kedmy, Israeli Illustrator, described his view of the education process in schools in this illustration: a place where the job of the teacher is to mold all students to be the same, while the job of the student is to succumb.

It's as if the vast majority of the world instinctively decided to place all the children wherever they are in little boxes just the same.[3] (see Fig. 1) In general, students are one big mass that needs to be fed. They are expected to fit into a uniform structure of time, space and information, most of which was defined back in the 19th century. They are still asked to memorize facts and figures, then throw everything back via tests. And more tests. Endless tests. We even have *global standardization* tests, not very far from the early-1900s IQ tests, reading comprehension, logic and mathematics skills, described in Chapter 3.

In view of the society we live in, students are required to develop skills that are likely to be outdated by the time they begin operating in what is known as the "real" world. We prepare them for jobs that by the time they finish high school may no longer exist, or at least change dramatically. We hardly stop to ask if job preparation should still be one of the goals of education. Even when attempts are made

[3] Paraphrase of the 1963 song **Little Boxes**, in which American singer Pete Seeger describes people all living according to the standards, means and goals that are all "just the same", with no room for creativity or uniqueness.

to "teach skills instead of facts", or as the latest bon ton claims "to develop 21st-century skills", these are usually universally defined skills fitting, at best, the near and foreseeable future. More times than less, they are skills that certain interest groups decided are necessary for society at large.

Where is our growing recognition of the diversity that exists between individuals and cultures within our species? What kind of innovative opportunities and unique features are we allowing our schools to offer? Are coping with uncertainty and developing agile flexibility part of our education processes? How much attention are we really placing on exploring with our students the *why, which and how* of what they are learning?

One would expect that an intelligent and design-loving species such as we are would utilize all the richness we are capable of wisely when dealing with grooming our young.

Instead, during the past 200 years, human society has changed dramatically. Schools have not.

The Best of Us

It's the first teacher training session. Some twenty teachers are in the class. After introductions, I ask each of them, "What exactly do you do in class?"

Twenty teachers, one after the other, will answer as follows:

"I teach chemistry"

"I teach mathematics"

"I teach mathematics and computer science"

"I teach agriculture"

And so on.

"That's nice, but inaccurate," I say, and forty eyes become confused. "You don't teach *subjects.* You teach *people.*"

If we agree that teachers don't teach subjects, rather they teach people, then we can continue and agree that teachers are not vessels of information or even skills. They, too, are people.

History's greatest teachers were people who had something meaningful and new to say about their areas of expertise, ideas that made them interesting to learn from and a source of inspiration. Students come to meet Nobel laureates to learn *their perspectives* about their research, not to "learn subjects" or hear a summary of their discoveries, which are all easily accessible in a variety of sources. They come to listen to astronauts to learn from *their experiences*, not to learn textbook or experimental astrophysics from them. They participate in master classes taught by leading musicians to learn from *their interpretations*, not how to read and play notes.

This is far from being a new idea. Suffice to remember that students came to Socrates not to study "the subject of philosophy", but to hear what *he* had to say about it.

Like learning, teaching is first and foremost a personal process. It includes the teachers' perspective, enabling the students to think about, discuss, approach and change with their own ideas. These are processes that develop attitudes, values, norms and skills. It's our responsibility as teachers to decide if what we impart will be the ingredients of a liberal, tolerant and democratic process, or something else. This rightly places the necessity of high-level professionalism and obligation on the shoulders of anyone who aspires to become a teacher.

This vision was expressed in the 1980s by Lee Iacocca, American automobile executive, who was quoted as having said,

> In a completely rational society, the best of us would aspire to be teachers and the rest of us would have to settle for something less, because passing civilization along from one generation to the next ought to be the highest honor and the highest responsibility anyone could have.

The best of us will be teachers, the rest will have to settle for less. Is there even one teacher training system that we know of that screens its applicants according to this vision?

One of the obstacles in our way to creating schools led by the best of us, who know how to educate in ways that are best for us, is that we haven't yet decided what that "best for us" should be.

> Mankind are not equally agreed about the things to be taught: should education be intellectual or moral? Should the usefulness in life, or should virtue, or should knowledge of the higher sort, be our aim?

These questions, articulated by Aristotle more than 2,400 years ago, are still being asked. They represent the various paradigms we hold regarding what education really is and requires.

Not the ones we *say* we do. The ones we *actually* do.

It's time to admit that education is not really considered a realm requiring high-level expertise. Most education systems don't screen, train or pay teachers as if they were one of the most important professionals to admire. Even their working environments aren't appealing or as respectful toward them as they should be.

We hold an implicit notion that we are all educators, simply because we all educate our own children. Most of us are convinced that we know exactly how schools should operate and claim that if only we were in charge of the education system, it would be excellent. Try arguing with parents who use the success of their children as evidence of their own expertise in educating others, and you will find it's a lost argument from the start. Unfortunately, the fact that I know how to successfully address a wound and which medicine is needed to deal with headaches doesn't make me a professional medical doctor, no more than knowing how to cook excellent food for my family makes me a professional chef. Personal experience, albeit often excellent, is not professional expertise, and success in familial or limited circumstances isn't evidence of success in large, systemic organizations.

Education is further mishandled because many of the actual policy makers for education are usually professionals in *other* areas. Professionals such as statisticians who work with numbers and graphs, management experts who look for cost-effective outcomes, high-tech and industry whizzes who think about their workforce needs, politicians who survive by being reelected, and other *not*-education professionals. The decision makers are too often "Masters of *other* trades" who have never actually taught children in formal settings for even one full year in their lives.

At least one full year in a classroom. A full academic year allows real experience in what it means to plan, implement and evaluate a curriculum as an ongoing, formative process. It necessitates fitting to the variety of capabilities, diverse needs, emotions, backgrounds and so on of twenty odd students in each class, day in and day out, week after week, month after month, through holidays and seasons, joy and crisis, for a whole year. Anyone who wishes to begin being considered as an expert educator or policy maker for our children must experience it.

Last but not least, while there is an abundance of international protocols, white papers, strategic programs, testing, academic research, facts and figures about what education should be, the practice itself is falling behind. In most countries, becoming a teacher is an extremely easy process, requiring a minimal screening procedure at best. A teacher is implicitly considered as someone who is there first of all to make sure that our kids have something, hopefully meaningful, to do while we are at work. While we are all familiar with outstanding examples, these are unfortunately not the ruling case. Under-funded school teachers and classrooms and outdated methods and tools are still the widespread situation of education.

Admittedly, frustration with our education systems isn't new. This is revealed when reading Charles Dickens's description of the flawed, contemptuous and abusive schools of his time; or in Lewis Carroll's 1865 "Alice's Adventures in Wonderland", in the wonderfully sarcastic conversation Alice has with the mock turtle, gryphon and others about school. Carroll declares what he really thinks about this institution in the words that he places in the gryphon's delicious testament:

> 'That's the reason they're called lessons,' the Gryphon remarked: 'because they *lessen* from day to day.'

At approximately the same time, Mark Twain's famous quote "I have never let my schooling interfere with my education" was only one of several critical remarks he had toward schools.

Almost two centuries later, we are still frustrated, feeling that our lessons have "lessened" and that the connection between schooling and education is remote.

Instead of using our assortment of competencies to continuously design and develop our means of inter-generational transfer, we are stuck in the same box over and over again.

Our education systems are fearful of leaving Plato's cave.[4]

Making Tomorrow Sweet

As I was travelling through Ethiopia, the scenes I saw in Vietnam, Ecuador and other developing countries repeated themselves over both time and geography: Children running after me for some sweets or money for their parents, who preferred their children stay at home and help the family by bringing additional income than go to school.

When a boy who looked no more than 12 years old tugged at my sleeve, I instinctively reached for some candy. He refused, shook his head and summoned me to follow him into his village. Curious, I followed, and his eyes began to shine as we approached a small hut. I was surprised to discover it was a small shop that contained books, stationary and writing materials.

He then turned to me, introduced himself as Johannes and said he didn't want candy or money. He asked if I could buy him a book. When he saw my surprise, he smiled and said, "Candy makes *now* sweet. A book will make *tomorrow* sweet." In other words, what he wanted was a future — and that meant education.

---

[4] **The Allegory of Plato's Cave** — an allegory presented by the Greek philosopher Plato, discussing our misguided perceptions of reality, our reluctance to change it and how it affects us.

In 1943, American artist Norman Rockwell described the meaning of the future in a thought-provoking manner. He portrayed a mother and father putting their children to sleep in their beds in the USA. As they tuck them in, the father is holding a newspaper and the headline announces the horror of the blitz on London. With it, the message of this serene scene becomes horrific as well: While these parents are calm as they watch their children drift safely into slumber, on the other side of the ocean parents are not so lucky, as they fear that their children might not see the light of the following day.

That, Rockwell claims, is our most innate, distinctive fear: That our children will come in harm's way. In his "freedom from fear" drawing,[5] he sharply claims that this is the one, most important desire we all have: A safe future for our children.

This isn't a new concept. In the Jewish faith, as we tell the story of Passover and the freeing of the Israelites from Egypt, we ask why ten plagues were necessary before the Pharaoh finally submitted to Moses' request, "let my people go".

Obviously, we explain, each of the first nine plagues was awful enough, and yet only after the tenth the Pharaoh finally yields. Why?

This is because of the concept that, as devastating as the first nine were — blood, lice, death of livestock, locust, darkness and so on — none of them was anywhere near as disastrous as the tenth: The death of all firstborn children. **While the first nine affect the here and now, the tenth is about destroying the future.**

For homo sapiens, securing the future isn't only about reproduction and physical survival: It's about the continuous expression of humanity as a unique species. For that, we have education. This realization inevitably leads us to understand that **the basic right of every individual is to be educated.**

It's time to take education back to school.

---

[5] **The Four Freedoms** oil paintings by Norman Rockwell are on display at the Rockwell Museum in Stockbridge, Massachusetts.

An Educated Being[6]

If ever there was a task of extreme complexity and potential, restructuring our education systems is the one. We will begin by considering the three dimensions I have mentioned — the personal, cultural and species contexts — and continue with the major questions that come to mind.

A personalized dimension of the educated self:
The most meaningful lessons I learned are related to personal experiences. They required the use of the one thing I know about myself better than anyone else and that none of my best teachers can know about me: How *I* learn.

Whether we are dedicating our time and effort to learn how to play the violin, speak a foreign language, solve problems in trigonometry or write a poem, our awareness of ourselves should be considered if we wish to succeed.

Questions such as how, when and where do I learn best? What do I already know? What interests me, what are my strengths and weaknesses, how do I approach something that I should learn even if I am not sure why? What can help me, what might disturb me? These are all questions that classrooms don't address, to an extent that, in many countries, a growing percentage of students don't "fit in".

These questions are important *for teachers as well*: What do I enjoy about teaching? How and when do I teach best? What do I need to be a better teacher? In a world where no one can know everything, how do I really feel when my students ask me something I don't know, or when they know more than I do? When they fail, how does

---

[6]The scope of this chapter is to address aspects of education as an activity unique to humans. I will not detail *solutions* to the problem called education, which deserve the attention of a separate book. Here, I will settle with presenting major issues and directions for thought.

it impact me, my sense of professionalism and confidence? Teachers should be able to enjoy support and guidance with these personal aspects of their profession.

Bringing these questions into the discussion of education is a complex task. Catering to every student's personal needs is daunting, time-consuming and requires growing resources, attention and high-level expertise. It's expensive. While we are open to personalized practice in a variety of areas, such as medicine, commerce and even communication and media, in most places the idea of personalized education is lagging behind.

Hope for change can be found in an unexpected place; if there is an upside to the COVID-19 crisis, it's what quietly developed in the realms of education by sheer necessity. With social distancing and quarantining paramount, schools shut down, reopened and shut down again. Teachers and students were literally and forcibly stuck at home for months at a time, trying to figure out how to maintain some form of education. The result is a variety of new creative educational models that brought into consideration the personal needs, tendencies and characters of teachers and students alike.

It is still too early to assess the long-term effect of the pandemic on education. Thus far, it has served to amplify a simple principle: When the system isn't right for too many of its components, instead of changing the students, it's time to change the system.

A contextual relationship with society and culture:
Six hundred pairs of eyes were staring at me. I was the honorary speaker of a national education conference in Seoul, South Korea, and the only foreign one. I was aware that the Korean culture considers speakers — be them teachers, lecturers or anyone else — as authoritative figures that should be silently listened to and obeyed.

Furthermore, I knew it is considered disrespectful to ask questions. So, I began my presentation by saying the following: "In my culture, at least once a year during the holiday of Passover, we place our 3-year-olds on a chair in front of the entire family, and encourage them to ask questions about the holiday. We expect them to continue asking questions everywhere, about anything, from anyone. In other words, where I come from, it is considered disrespectful to *refrain* from asking me questions, since it might mean that I wasn't interesting enough. So please, feel free at the end of the conference to ask me any question you might have."

I finished my presentation, as did the other local presenters. The MC then asked the audience if they have any questions or remarks to any of us. There were five questions from the audience, all of them addressing me alone. It wasn't because my presentation was the only thought-provoking one. It was because I was the only presenter that allowed questions.

Education is probably one of the most ancient behaviors of homo sapiens, stemming from the evolutionary development that led humans to organize in groups, hence to be born and live within the context of specific cultures. Created by us, our cultures are developed by our loyalty to narratives, beliefs, values, ethics, attitudes, politics and rules. Once we organize our lives in accordance with the governance, demands and expectations of a culture, they become real differences, creating both richness of diversity as well as separation.

Albeit our cultures are man-made and can therefore be changed by humans, we tend to adhere to them relentlessly, stubbornly resisting change. This, too, isn't surprising: Our cultures give us a sense of belonging as well as confidence, a sense that we fear to break and lose by change. "Without our traditions, our lives would be as shaky as a fiddler on the roof,"[7] sings Tuvia at the beginning of the famous musical, only to spend the entire show dealing with the conflict between the traditions that build the culture he belongs to and change.

Almost all education systems and processes act upon the cultural dimension. Considering that practically every aspect of our lives is a result of the culture we belong to, it's easy to understand why. Language, religion, gender equality, holidays, historical narratives, customs, arts, governance, etc., are all based upon cultural perspectives and are strongly present in education systems.

---

[7] **Fiddler on the Roof** is a 1964 Broadway musical. Based on a book series written by Jewish Shalom Aleichem, who described Jewish life in small towns in Russia. Among other subjects, he described the inevitable tension of preserving a unique culture within a larger social context pressuring for assimilation.

Rethinking the cultural context in education is a difficult task. Despite the understanding that questioning and doubt are the source of development and richness, most cultures will avoid encouraging their young to ask basic questions about their own foundations, fearing that answers might lead to change of traditions, confrontation of fundamentals, loss of control, and eventually, disappearance. The challenge is to introduce it through the lens of respectful exploration, tolerance and acceptance for the sake of promoting healthy human relationships.

A universal concept of Homo Sapiens:
The urge to educate is shared by humans everywhere, regardless of geography or culture.

Homo sapiens share the traits of curiosity, eagerness to discover, explain, change and communicate to others. We enjoy playing, we share the inclination to imagine and create narratives, using a variety of skills and translating them into exciting actions and products, continuously learning from our failures as well as triumphs. It's part of what defines our species.

These universal human traits are a fountainhead of tools for creative, multifaceted education processes. Using them is the complete opposite of the fixation and duplication resulting from "This is not the way we do it". It leads to creativity, imagination and exploration, confident independence, growth and change.

If implementing these three dimensions isn't challenging enough for you, try adding basic, good old *wh-questions.* They clarify the extent of the complexity we are facing. Brace yourself for more questions than answers.

Who

Driving in the early evening, I suddenly felt confused. "I'm not sure this is the correct turn," I mumbled aloud, when my granddaughter raised her head from the back seat, her eyes left the screen she was looking at and she said in the impatient tone of a smart aleck teaching a clueless individual, "Grandma, just use Waze."

She was 3 years old.

It is time we honestly explore who our children are and to begin using the tools they understand and live with every day. The rapid changes human society has been experiencing during the past century hasn't skipped them, and children today are growing up in a world that their parents and teachers aren't really familiar with and don't always recognize. What do we know about them, what are the characteristics that they bring into the education process? What are their views, desires and interests? What do we understand about them and how can this understanding help us build with them a meaningful learning process?

The same questions must be honestly asked about the profile of the teachers we desire for them.

What

When I was in high school, we were required to choose a topic that interested us, explore it and write what was called a "personal paper", a term coined by the principal of the school. The topic, we were told, can be about anything we want, as long as we use multiple resources to explore it and find at least one of the teachers who would be willing to be our mentor throughout the process. It was an assignment that was unique to the school I attended and compulsory for all students at that specific school. Although it was an additional burden to everything else the state schooling system demanded, most of us approached it with enthusiasm: It was a rare opportunity to do something at school that was of our own choice.

I decided to explore a question regarding the influence of singalongs, an Israeli phenomenon that began long before the establishment of the state and developed into an accepted, popular social pastime. I wanted to understand how the songs we sing represent and influence the development of Israeli society and culture. I was a sing-along lover, and found it interesting. My history and music teachers agreed to mentor the process, and I began to work.

For 3 months, I explored. I spoke to song writers and interviewed social studies and history experts. I learned about poetry genres from literature experts. I visited a firm that dealt with marketing and branding, an area that I was just beginning to be aware of. I read anything I could find that would help me address the different angles I was interested in, including music sheets, diving into structure, rhythm, meaning and connections to musical instruments. I needed to learn and understand the historical trends and social changes of the 19th and 20th centuries, including the geopolitics of the time. I had to explore the renewal of customs, including the everyday use of the Hebrew language, how semantics and symbols developed. I had to understand how words and music can be used to create social structure, enhancing motivation, devotion and community building. Under the guidance of my teachers, I found even more angles that we discussed and decided should be further visited.

To this very day, almost half a century later, I have very little recollection of the official subjects I studied in class, those that the state decided were necessary for my future. On the other hand, I can still discuss in depth my "personal paper". It was based on my choice of interest, which sparked my motivation to explore into the engine of what became a fascinating and enriching process. It sent me on a journey that, more than about the subject itself, taught me about myself, my abilities and helped me develop skills that became valuable assets for life.

Our ability to generate and mold knowledge to fit our needs has become almost endless. This advantage comes with a price: The abbreviations "TMI" and "TL;DR"[8] have become daily visitors in our lives, and for good reason: There is absolutely no way we can grasp the flood of growing information we are drowning in.

We need to revisit the question of "What is the 'what' that we must teach" and what are the best ways to introduce it in a balanced, exciting system.

---

[8] **TMI** and **TL; DR** — abbreviations for **Too Much Information** and **Too Long; Didn't Read**. Both represent the shortening attention span we are developing toward information presented in long texts or other forms that require spending more time and attention than we are willing to invest.

How

"*Go!*"

At 6 years old, I was sitting on a bicycle without training wheels for the first time. I could feel the excitement and fear swirling in my stomach as my father pushed me forward, encouraging me to ride. Being used to the sensation of flying on the long pavement that stretched in front of our house with training wheels, my legs did what they were used to. It worked for a few seconds, after which I found myself flying into a large, thick rose bush. The thorns and insult stung hard, but instead of rushing to embrace and console me as I expected, my father just stood there and calmly exclaimed,

"Get up and try again."

I looked at him with tears.

He wasn't impressed.

"Get back on the bike and try again," he repeated in a calm voice.

Grunting, I did as I was told. I tried again, this time trying to lean forward as I rode.

It didn't work.

After two more attempts, I finally succeeded in keeping my balance and rode across the entire pavement without falling.

My father, still calm, just said, "Good for you. Now you will remember that whenever you fail, it means you need to get up and try again until you get it right."

"But what if I can't?" I asked, holding back a few tears. The thorns still stung.

"You can," was all he said. "Learn and try again. That's all."

And, with that, we went to search for some Band-Aids.

Get up and try again. One of the best lessons I had learned.

Anyone who has watched children in a playground can observe the following: They are trying things, failing and trying again. They are inventing and enacting stories, switching roles, building castles and fighting dragons. They make decisions, solve problems, role play, build and invent.

As they grow, the processes will be more connected to their relevant, everyday lives — a tattoo will replace a costume, social connections and rivalries will replace castles and dragons, but the processes are similar: Experimenting, stories, problems and decisions, repeating themselves in endless spirals. It is the kingdom of the big five discussed in Chapter 2, enabling us to clarify values, understand who we are, what we want to be, how to get there and develop both our individual and social characters.

This rich, active, multidisciplinary process is minute in schools. We introduce the realms and riches of human knowledge through disciplines. From first grade to universities, we teach in silos, with each discipline "minding its own business".

Worse still, most of the time, we *present* them. Hands-on activities, trial and error, exploration, outdoor activities, games, etc., are rare, usually connected to very specific areas or extra-creative teachers.

Reality isn't made of separation, and it isn't passive. Too many times, teachers are stunned when students don't necessarily make the connections and transfer from one discipline to the other. Learning how to effectively and responsibly connect and use information, while turning it into relevant, active knowledge, is a keystone of effective education.

How we educate is also in dire need of developing the art of interacting with ourselves and with others. If schools should have a social goal, this is the one: To enhance the conduction of positive, ethical and sustainable social interaction.

Where and when

People in India interacting in real time with people in Kenya and Brazil has been around since the turn of the 21st century. We no longer need to fly across oceans to have fruitful interactions with peers living on other continents. Time and space have dramatically changed since the World Wide Web became a trivial household entity.

If we agree that education is continuously constructed from *all* our experiences, the question of "where and when" is answered within. Until recently, using diverse time and space for educational purposes seemed like a far-reaching tool, unavailable to most. Then came the COVID-19 pandemic, forcing us to shutter the boundaries of the traditional classroom. Proving that while students and teachers still love going to school for social reasons, the act of "schooling" itself lost its limitations as kitchens, bathrooms, gardens, digital environments and virtually any place that children and teachers could use while under restrictions were indeed used.

Science was done at home, history was visualized by documentaries, literature was expanded beyond curricular restrictions. Time itself changed, as learning became explorations that were suddenly awarded new time slots and durations. More and more hybrid learning, merging both physical and digital arenas as part of the education process, was conducted. Hybrid learning's success is based on the ability to balance between environments, without creating "digital fatigue" on the one hand or adhering to physical boundaries on the other. If the only opportunities left from the pandemic are medical ones, without harnessing the educational triumphs that sprouted from need and new-found capabilities, COVID-19 will be a war well wasted.

Why

The fact that our education systems are built the way they were built for centuries isn't an excuse to continue them the way they are. In the midst of the most complex, enabling and global century humanity has ever known, our education systems are stuck with methods and structures that were mostly established during the Industrial Revolution. Considering we have since gone through what is known as the technological and information revolutions, schools are stuck at least two revolutions ago.

If we needed an additional "wake-up call" to begin questioning why we do what we are doing, along came COVID-19 and proved that our historical reasons are no longer relevant, nor can we continue to disregard the natural environment and our relationships with it. Why we organize our academics the way we do is a question we need to honestly explore, accompanied with the courage to "let go" and discontinue when the answer suffers from ulterior motives or just isn't good enough.

Falling Walls: A Quick Look at Academia

Finally, we will turn our attention to take a quick look at the role of one of the main sources of change in the past two centuries: *Our universities.*

I was CEO of the internationally acclaimed Wolf Prize,[9] awarded to leading scientists and artists from all over the world, when the Higgs boson particle[10] was discovered at the CERN particle physics laboratory near Geneva, Switzerland, in 2012. A short while after the discovery, I received a call from one of the physicists who worked on the project.

"I think we should receive the Wolf Prize in physics," he said.

"I agree," I said for the sake of argument, putting aside the fact that the prizes are always given by professional international committees which I rightfully had no influence over whatsoever. "But who should it be given to?" My question was an honest one. Three thousand scientists from all over the world from different universities, organizations and disciplines worked on the project. The prize would need to be given to, well, humanity at large.

One of the major processes of this century is the dissolving of traditional boundaries. Zones that have traditionally been separate and compartmental are integrating. Borders between nations

[9] **The Wolf Prize** was established in Israel by Ricardo Wolf in 1981. It is a merit-based award, given to excellent scientists and artists regardless of nationality, gender, religion or age, and is internationally considered second only to the Nobel Prize.

[10] **The Higgs boson Particle** is the particle that gives all other fundamental particles mass, according to the standard model of particle physics. Long sought after, its discovery is considered as a significant breakthrough in theoretical physics.

and cultures, between physical and virtual spaces, traditional and liberal approaches, formal and informal settings, private and public arenas, teamwork and individual achievements and many other realms are quickly merging. Communities are based more on interest groups than on geography, as was displayed by the latest and most rapid example of the race after the COVID-19 vaccination.

During the 20th century, you might still be concerned about time and space leading to ***brain drain*** — the relocation of educated, highly professional and productive people from the country you live in to another country for long periods of time, thus "draining" the human capital available and necessary for your society's growth.

The reality of the 21st century turned it into ***brain circulation*** — building on the continuous mobility and relocation of people to and from places, circling around, bringing and taking their expertise with them, reducing the limitations of time and space.

This, too, is now quickly being replaced with ***brain integration***: Global communication and transportation has developed to an extent that allows people to stay put while still extensively cooperating with experts and resources from all over the world, regardless of location, almost completely dismissing the importance of time and space.

As a result, the world has become an open candy box for us to enjoy. Many of us can work anywhere, at any time. We are exposed to the richness of humanity at large and to environments that not so long ago were but a dream of risky travelers. Intellectual growth, research and creative development have become hybrid, combining physical and digital arenas and supported by unprecedented resources invested by governments, policy makers, industries and the public.

Lo and behold, our universities are like our schools: Falling behind while refusing to accommodate and adapt to our ever-changing world. On the one hand, they are the legitimate hub of philosophy, intellectual growth, creativity and basic research. On the other hand, costs of academic studies are increasingly rising, creating a major financial challenge for college students. Mostly for these financial reasons, universities have become more and more vocational prone. The outcome is a growing gap between the traditional world of universities and the rapid changes of reality.

This has led to the development of alternative initiatives of study and research, created and offered to young adults by non-academic institutions and organizations. They are making good use of the world now being at our fingertips, challenging the traditional world of academia. Large multinational companies, such as Google, Microsoft, Disney and many others, have created their own "higher education schools". Enticing young adults to abandon the traditional path to full academic degrees, they present them with a variety of alternative paths, from free massive open online courses to "nano-degrees" and focused executive programs. The companies involved collaborate in the process, accrediting and recognizing each other's courses, making them more attractive and relevant. More and more young adults are choosing these and other innovative, untraditional paths that are more fit to their needs and allow them at the same time to work and gain experience and money within the companies involved.

Although there are initiatives to create different models of study in academia,[11] these are less than a handful. If the academic world aspires to continue to be a relevant hub of creativity, research, development and intellectual growth, it needs to ask the same questions other parts of the education world should ask.

Our education systems are still the best structure we have to disseminate knowledge, and the fact that education is compulsory by law in almost all countries in the world places human knowledge as a democratic right for all. In this regard, the education systems we established in the 19th century are some of our greatest triumphs.

---

[11] Quest University in Canada and Minerva-KGI in the USA are two examples.

On the other hand, the growing gap between the education they offer and the pace of change and growth of human needs in every other aspect of life is still staggering.

***

A full discussion regarding the aspects of education, theoretically and practically, is yet to be developed. I will end this chapter with a vision that is relevant for our current journey.

Imagine this: An education system that is people and curiosity centered. A classroom that is vibrant. Students and teachers who are absorbed in immersive learning, led by agile choices and interactions. Disciplines aren't taught as silos, but are merged together to be explored and developed as they so naturally do outside the classroom. The world is the classroom, integrating between opportunities and environments. Students and educators are inspired to envision, explore, design and impact. Together, they are a system that has faith in itself and its ability to engage in independent, ethical critical thinking, doing so in responsible, sustainable ways for the benefit of their own community and the world at large.

If anyone can achieve this vision, it is us: The species that knows how to lead itself from imagining to exploring, understanding and doing. It is what has enabled us to develop the exceptional paradigms and practicalities of the relatively recent development of homo sapiens — the game-changing area we will discuss in the final chapter: ***Modern science.***

Chapter 6
Everything is Science*

Like in every other place in the world, it was easy to identify the schools in Uganda. Long, single-floored buildings, with several doors usually facing a yard. In Africa, many schools will use the outside walls as resources of information, covering them with drawings — of the solar system, maps of the world, diagrams of human bodies, and so on (Photos 108–110, pp. 217–218). Messages are scattered throughout the schoolyards: "Absenteeism leads to failure", or "Stay in school, complete at least primary seven" and even "I supported my daughter to complete her education. Now she is a medical doctor". In a continent where a substantial percentage of parents in rural areas prefer to send their children to work instead of to school, these come as no surprise — it's a matter of day-to-day survival.

The school in the remote village I visited in Uganda was no different from the ones I visited in Ethiopia, Madagascar, Botswana, Kenya and other African countries, with one exception: This was a spontaneous visit that took place during the COVID-19 pandemic.

As I approached, the principal of the school, a smiling woman named Pamela, greeted me. Although I was an unexpected guest,

---

*Contrary to the protagonist's sentence "I don't think anything is ever just science" — from the novel "**What I loved**", by Siri Hustvedt.

her "Welcome!" was genuine and displayed her perfect English. Pamela introduced herself and told me about the elementary school she headed of 520 students from grades 1 to 7. She then invited me into her office. Next to the entrance, I read the vision, mission and motto of the school (Photo 111, p. 219):

Vision: producing self-reliant and God fearing citizens;
Mission: to provide basic education;
Motto: Education brings success.

"I'm afraid I only have two classes to show you," Pamela said and started describing how they were coping with the health regulations of COVID-19. "We are teaching in shifts of 3 months," She explained. "Now we have grades six and seven, and the rest are on leave. In a month, grades one and two will come back, and grades six and seven will join grades three, four and five. So we rotate."

I noticed the 2020 timetable for grade seven on the wall. Beneath it, I could see the timetables for 2019 and 2018, before the pandemic had begun. They were very different. Pamela noticed my interest and explained, "Because of the need for smaller groups, we couldn't continue our regular schedule." She sounded a bit apologetic. "We had to choose which subjects were most important to continue, and those are the subjects the students are learning. Everything else will have to wait." When I asked how one chooses what is most important to teach, her voice became proud. "We chose the subjects that we believe are crucial for equipping our students with knowledge and tools that will ensure a better future for them and for our community," she answered. "It was easy. We insisted on science, math, social studies and English. These are subjects that can't wait for the pandemic to end. Everything else, such as music, literature, history — they are important, but will have to wait."

In the two classrooms that were operating, the sixth and seventh graders greeted me with enthusiasm. They wanted to know everything about the place I came from, about our schools and how we treat the pandemic. One of the classes was in the middle of a science lesson, studying all about germs and viruses that generate diseases — why, how and especially what to do to avoid them. The other class was busy with algebra.

"I am a science educator, so I understand why you insisted on science and math," I said to Pamela after we left the class. "But could you explain why you chose social studies and English as well? Why not something else?"

She smiled again. "Science and math must be viewed as a relevant part of everyday decisions," she said. "That's what social studies will teach them. English gives them the ability to be part of the rest of the world, or at least most of it. All put together, they will be better prepared to take responsibility for their future."

I couldn't agree with her more.

The Most Important Key on the Keyboard

Science is probably the most extreme result of combining our big five with our technological intelligence and inclination to educate.

It requires complex abilities, such as the following:

- The ability to imagine not only through stories, but through the use of facts, figures and evidence;
- The skills to create practical means of exploration and discovery;
- The expertise of communicating our understandings in ways that others can either repeat and develop further, or prove should be rejected;
- The capability to willingly defy and change what we may have previously concluded; and
- The talent of engaging others in our logic, convincing them that human society is better for it.

It is also one of the most difficult areas to explain and disseminate, suffering from continuous denial, rejection and even abuse, as I will soon discuss.

A result of our uniqueness as an organism, science is one of the greatest, exceptional achievements, exclusive to homo sapiens.

Science is the best way we have to understand reality, its resources and phenomena, using the results of our explorations to design and shape all components of our lives. Harnessing the information, methodologies and impact of science in all aspects of life is probably one of the most important tools we have to solve problems and make everyday decisions based on evidence and logic.

Stemming from curiosity, the development of science is led by ongoing questioning of all sorts: Why is something as it appears? How does it happen? What are the consequences? When, where and what can I make of it all?

In the scientific world, the answers are never the final point; rather, they serve as the substance leading to more questions, and then more.

The most important key on the scientist's keyboard is the question mark.

Turns out that there's nothing like "a good pandemic" to make this point clear. If anything good can be said about COVID-19 that hit the world in March 2020, this is it: Science was placed in the limelight. Reality was suddenly all about looking for logical, evidence-based answers to questions. People who would have nothing to do with science, who would frown at anything that resembled dealing with "tiresome figures, graphs, mathematics and all those strange and long-name definitions" claiming it was fit only for geeks, suddenly became experts in genetics, immunology and statistics. Viruses, antibodies, herd immunity and vaccinations became household words, relevant and discussed by everyone. Scientists, medical doctors, public health experts, epidemiologists and mathematicians became celebrities, sought after and placed in the focus of public interest. News, fake news, facts and rumors reached unprecedented heights of attention and debate.

If questions such as "What can I do about climate change?" don't always strike us as something directly connected to our behavior by immediate results, along came COVID-19 and placed the relevant connection between science and society at the front stage. Questions became personal, even intimate: Why do I need to wear a mask? What kind of hygiene rules can help prevent disease? How can

I help my elderly parents while distancing myself from them? If the virus is transmitted through my breath, what should I refrain from doing? Every question generated more questions.

The entire world was suddenly trying to learn science.

At the Davidson Institute — the educational arm of the Weizmann Institute that deals with science education, communication and dissemination and that I am proud to lead as CEO — we had unprecedented demands for answers and information from all realms of society, in hope that science will help understand and give rational tools to deal with the uncertainty and chaos of the global pandemic. "Never in my wildest dreams did I believe that a day would come when I would be asked to appear on national TV, during prime-time news, on a panel discussing mRNA!" a colleague of mine exclaimed ecstatically. Be it through digital media, TV, radio, social networks, etc., we discovered that every subject and angle of science that we offered, from articles to discussions, activities, courses, games, riddles, experiments and more, was devoured by millions. It was a delicious interpretation of Dickens's[1] words: One of the worst of times for human society — and one of the best of times for science.

Needless to say, this all goes way beyond and long before COVID-19 appeared. The history of science is extensively documented and discussed. Bare in mind that the science I am referring to here is ***modern science*** and especially the ***scientific methods*** that have rapidly developed since the Scientific Revolution of the 16th century. Since then, science has become our central way to push the boundaries of what is known, to relentlessly pursue our aching need to broaden our horizons of understanding of the natural world around us, working with our technological intelligence to experiment, develop and create new means for the benefit of our lives.

There are three main reasons that make science a crucial and inseparable part of human culture's ability to prosper, both on the individual and the social levels.

---

[1] Based on the phrase from "A Tale of Two Cities", an 1859 novel by Charles Dickens: "It was the best of times, it was the worst of times."

The first and probably most popularly understood is ***socio-economic***. Science is one of the main power engines of every modern society. From medicine and pharmacology to food tech, transportation, security, communication and more, it lies at the base of every single industry, fueling the development of creative and growing production. A culture that continuously develops sciences, as well as the arts and skills connected to them, such as design, media arts, technology and so on, strengthens its sustainability. A culture that waives them critically impairs its ability to develop state-of-the-art innovative systems and products promoting health, employment, security, leisure and education. Without leading experts in sciences, it is impossible to maintain a leading, modern economy and culture in the 21st century.

The second reason is an ***ethical*** one. Since the 16th century, we have accumulated levels of scientific and technological understanding, know-how and achievements that allow us advantages we have never had before — as well as unprecedented power to cause harm at scales that we must address using responsible, rational skills. These cannot be fully expressed without having basic knowledge and understanding of science, technology, their relevance and interactions with the environment and human societies. In a democratic society, this means the right of basic science education for all.

The 21st century emphasized the need to be able to implement knowledge and skills to new situations, many times unexpected. This places the scientific method in a crucial position for the continuing development of an ethical, just and sustainable culture. It raises questions such as these: How should eligibility for organ donations be decided? In which cases is genetic diagnosis recommended before pregnancies? Who is responsible for the food security of the underprivileged? Why is green architecture so important? Why must we invest in alternative energies? These are some of the issues that require value clarification and ethical discussions to deal with attitudes, beliefs and emotional relationships, to be able to contain the outcomes of our decisions.

One of the most important questions we face is "How do we decide that the fact that we *can* do something, means that we *should* do it?" The answers we give to this question aren't based on science and technology alone; rather, they're based on their connection to society, its norms and its ethics. Without understanding the centrality of science and technology in our lives, our ability to deal with the ethical, social and environmental aspects necessary to build a just, pluralistic, tolerant and democratic society might be left in the hands of charlatans and interest groups, such as political powers, religious groups or industry leaders, creating fertile soil for manipulation and hidden agendas to flourish.

The third and probably most neglected reason is that ***science is one of the characteristics that make us human.*** The scientific method is naturally displayed in our behavior, as demonstrated by every baby throwing an object again and again on the ground just to watch it fall and hear the sound it makes, by every toddler asking why birds sing, clouds move and leaves fall, expanding all the way to the scientists asking what are the smallest particles of matter.

Science and technology are a striking display of what the unique characteristics of homo sapiens can achieve. They are the devices we have developed so we can dive into the deepest reefs, fly over forests, reach the moon and beyond — metaphorically as well as literally. They are qualities that need nurturing from a young age on, persistently turning each area into an adventure of discovery and creation.

The last two centuries have proven that we reach the peaks of our abilities through the "marriage" between science, technology and society — such as reflected in the light bulb that harnessed science and technology together to free us from our dependence on sunlight, enabling us to radically change the time and space of our activities, or in what the combination of science and technology of radio waves did to our social ability to communicate around the globe, be it through television, cellphones or the World Wide Web; the list of examples is endless and multifaceted, including exciting and meaningful ways to promote our lives as well as dark, harmful outcomes when dealt with recklessly, as I will discuss shortly.

No matter how we look at it, science and technology are made by humans, for humans. They are a product of our ability to think, create, try, fail and succeed. They are central to the works of every human society, and should therefore be accessible, available and understood by everyone. This is especially true if we want them to be used in accountable, sustainable ways.

The best way I know to achieve this goal is through ***science education for all, at all ages, all levels, via a bountiful variety of relevant, innovative and creative experiences*** — a strategy that I have dedicated my professional life to.

Let's take a few steps to understand some of its meanings.

A Complete Mind[2]

> *To develop a complete mind: Study the science of art; Study the art of science. Learn how to see. Realize that everything connects to everything else. (*Leonardo da Vinci*)*

It was an age of relentless curiosity, exploration, discovery and invention. The printing press was a new development. New lands were discovered by Europeans, bringing new opportunity as well as clash of cultures, exploitation and cruelty. Ancient ideas were revived and questioned, while new ideas were introduced. It was the age of the Renaissance in Europe, an age that allowed celebration of humanity, beauty, philosophy and creativity.

It was during this age that Leonardo da Vinci not only defined the crucial importance of bringing the sciences and liberal arts together but put this idea to practice. His entire portfolio — sketches, notes, paintings, frescos and more — is an ingenious combination of art, technology, botany, geology, anatomy, optics, architecture, color, mathematics and so much more. This striking array of diversity is evidence of the heights of originality and confidence that the use of combined, multidisciplinary knowledge and practice can achieve. Using his imagination, creativity and ability to symbolize and communicate his ideas in ways that we can all understand and follow even 500 years later, da Vinci was a man of unsaturated, unlimited and universal curiosity. Half a millennium after his death, we are still trying to find adequate ways to follow his

[2] "The Complete Mind" is a concept developed by Leonardo da Vinci, 16th century.

call to develop "complete minds" by exploring, recognizing and celebrating the multifaceted interconnectedness between all aspects of the reality we live in.

Developing "complete minds" requires a combinatorial, multidisciplinary, complex systems approach, crucial for the success of individuals and society as a whole. Graduates of the sciences (mathematicians, biotechnologists, engineers), the humanities (politicians, archeologists, philosophers) and the arts (musicians, designers, choreographers) all depend on one another. Culture grows from the combination of their actions and achievements. A society that underappreciates one will severely disable the other. A mathematician without the understanding of poetry, an author without environmental awareness, a computer engineer without knowledge of music and a philosopher who hasn't studied evolution — all may be achievers of excellence in their fields, but the society they will build will be a community of narrow-minded individuals who find it difficult to communicate with each other, greatly lacking in pluralism, tolerance and democracy.

As the 20th century developed, it became evident that the achievements as well as challenges we face are fundamentally intertwined with the growing connections between scientific, technological and social abilities. Actively changing the environment at large as well as our place in it, our species has become the central, most impactful organism on the planet. This status comes with responsibility, emphasizing the necessity of multidisciplinary logical, evidence-based decision making and sustainable problem solving. As a result, since the second half of the 20th century, we are witness to a new definition: ***Scientific literacy***.

Scientific Literacy: Getting to Know Us[3]

"I can't stand science anymore!" my daughter stormed in from school one day. She threw her backpack on the floor, obviously frus-

[3] "Getting to know you" is a song from the 1951 Rodgers and Hammerstein musical "**The King and I**". The song praises the advantages of a warm and affectionate relationship between different sides.

trated. "What a dull lesson we had today! I don't understand why on earth I must learn these things!"

Keeping my calm and holding my breath, I gently asked, "OK, maybe you don't need to learn science. What is the subject you are studying?"

She looked at me, miserably. "Reproduction," she said. "Something about cows, chicken and the hormones of ovulation, something like that. *Boooooring*!"

It was my turn to be miserable. There I was, standing in front of my then 14-year-old adolescent, in the prime of her youth, filled with bursting hormones herself and just beginning to discover the realm of sexuality, reproduction and its implications on her life, and her science lessons are killing any sense of curiosity she might have about *the* most relevant topic to her current world.

She was right, of course. Shying away from any kind of creative involvement or engagement of the students, their interests or the connection and practical relevance of the subject to their lives, her science teacher meticulously followed the curriculum designed by the state. It included facts and figures about the reproduction systems of cows, chicken and other organisms, including at some point the graphs of hormones that are involved in human reproduction. Important as they may seem to me as a biology enthusiast, for her and her peers, it was just a pile of facts and figures that had to be memorized for an upcoming test, after which they can all be forgotten.

Licking my professional wounds, I somehow managed to convince her to embark with me on a process of correcting her impression of the subject of reproduction. It was simply one of those subjects that we all need to basically understand if we wish to be able to function responsibly and successfully in our lives.

It's being scientifically literate.

Traditionally, the term "literacy" has been used to describe a person's ability to read, write and perform basic arithmetic functions, namely, addition, subtraction, multiplication and division. In the world post World War II, it became evident that these were not enough to consider a citizen a literate person.

As new disciplines emerged and developed, knowledge and performance became more sophisticated and awareness of contextual paradigms grew. It became obvious that a literate person must be able to incorporate knowledge, understanding and know-how in operative, flexible and contextually relevant ways in order to solve everyday problems and make decisions in a rapidly changing reality. It became widely accepted that reading, writing, numeracy and even basic declarative knowledge are not enough to be considered a literate citizen of the 20th century. In the 1950s, UNESCO defined the term **functional literacy** — the level of literacy that one needs to successfully function in the social context within which he or she lives.

By definition, functional literacy is personal, context-dependent and dynamic. It involves cognitive, affective and social processes. In each field, it is a continuum that develops and can be learned from the moment we are born.

In each field, we will develop at a different pace, level and time. We can be highly literate in history, while less literate in mathematics and maybe somewhat literate in music. We can be extremely literate in sports, while developing a mediocre level of literacy in art. In each area, functional literacy includes the array of practical knowledge and skills relevant to that area and its connections with other disciplines, necessary to perform holistically, effectively and successfully in an ever-changing, dynamic and multifaceted environment, in real-life situations.[4]

Functional literacy is the cement necessary to ensure that our society operates stably. It is a tool for development, enabling social mobility and equal opportunity. It is the basis of commonality in shared fields, enabling a plane to fly from one part of the globe to the other, a company to thrive within forces of global economy, a community to build sustainable water and food supply in a desert,

[4] Charts measuring and comparing literacy in a variety of areas, age groups and different countries of the world can be easily found on the Internet. Obviously, due to changes in educational goals and subsequent reforms that different countries implement, they change periodically.

and artists to create cross-cultural insights. Without it, dealing with the challenges we face will be impossible.

Functional literacy isn't limited to the ability to read and write in a specific language. It's the ability to know and use a variety of "languages" in the widest socio-cultural way. **One of These Central, Most Relevant "Languages" is Science**.

Stemming from the concept of functional literacy, **scientific literacy** is the basic level of understanding and know-how in science that is needed for every citizen to adequately perform in a society driven by science and technology. Regardless of our profession, age or socio-economic status, we all make daily choices that the lack of scientific literacy critically impairs our ability to deal with, as COVID-19 so amply proved.

Furthermore, with the scientific community's growing need for investment of meaningful resources to develop new medicine, innovative and advanced agriculture, communication, transportation and so on, scientific literacy is also a central tool needed for the public to appreciate and support the work of the scientific community, understanding its relevance and importance in the daily realities of each and every individual.

The call for scientific literacy quickly found its way into education systems in the 1980s, from elementary school to academia. Numerous ideas were developed and implemented, in the US, Canada, Europe, Israel, etc. Paradigms chased paradigms, reforms[5] replaced reforms, acronyms were changed to other acronyms, national and international tests and assessment tools were developed and implemented, all seeking to achieve one major, difficult goal: The development of scientific literacy for all as a tool for a better, advanced human society at large. Unfortunately, regardless

---

[5] To name a few: Project 2061: Science for all Americans; STS — emphasizing the connections between science, technology and society; STEM — emphasizing knowledge and skills in science, technology, engineering and mathematics; International science education projects and especially assessments run by the OECD, the UN, the IAE — such as TIMSS and Pisa.

of these reforms, the level of scientific understanding has not sufficiently changed, and its place in society is still in aching need of improvement.

The challenges that achieving scientific literacy faces are numerous. One of them is the complex task of **staying in sync** with the advances and innovations in science and technology as part of the education system. Another is the task of **communicating science** in an understandable, fit-for-all fashion. A third, probably one of the most intriguing, is the **affective** challenge: The attitudes, social paradigms, preconceptions and feelings we have toward science have a significant impact on our motivation and even on our ability to achieve scientific literacy.

Of the various challenges, I will highlight two major, unsolved issues: The steadfast, long and ongoing position of **science denial** and the need to **adapt the development of scientific literacy to an agile, boundary-breaking and swiftly changing reality**.

Science Denial: Let's Call the Whole Thing Off[6]

In 1829, Edgar Allan Poe wrote the following sonnet:

Sonnet: To Science/Edgar Allan Poe

Science! true daughter of Old Time thou art!
Who alterest all things with thy peering eyes.
Why preyest thou thus upon the poet's heart,
Vulture, whose wings are dull realities?
How should he love thee? or how deem thee wise,
Who wouldst not leave him in his wandering
To seek for treasure in the jeweled skies,
Albeit he soared with an undaunted wing?
Hast thou not dragged Diana from her car,
And driven the Hamadryad from the wood
To seek a shelter in some happier star?

[6] **"Let's call the Whole Thing Off"**, a 1937 song by George and Ira Gershwin from the film **"Shall We Dance"**.

> Hast thou not torn the Naiad from her flood,
> The Elfin from the green grass, and from me
> The summer dream beneath the tamarind tree?

Poe laments over the strong dichotomy between science and, well, everything else. Especially, what, in his eyes, makes our hearts celebrate and our spirits soar. Science, according to Poe, is a malicious predator. It preys on imagination and is an enemy of creativity and art. In his eyes, science is no more than a way to describe "dull realities". Science cannot be loved or regarded as wise or beautiful. It is not for the poet's heart, nor for anyone seeking excitement, realization of dreams or even happiness. The sonnet is a poetic, rhetoric case of denying that science can ever be a source of vision, thought, inspiration or magnificence.

Science denial has been embedded in human cultures for centuries. The example of Copernicus and Galileo's defiance of the traditional faith of the earth being the center of the universe is well known and documented, as is the objection to Darwin's Theory of Evolution. To this day, in spite of an abundance of evidence validating these schools of science, creationists are still plentiful. So are groups of people who, while they might reluctantly consent that the earth is round, still believe that it is a flat disc. Round, but flat.

History proves that opposition to scientific theories usually rises when they contradict belief systems, economic interests or political power. Opposition to Copernicus was weaker than to Galileo for reasons of rating: Copernicus published his heliocentric theory in Latin, the language of the scholars that only few could read. Galileo, on the other hand, published in Italian, the language of the masses. In Bertolt Brecht's play "Life of Galileo", when Galileo suggests that he will write his next work of science in the common language instead of the scholarly Latin so that it can be read by everyone, he is confronted with the idea that the noblemen cannot support science that challenges the Bible, since it would upset the peasants and damage their land's economic prosperity. The scene portrays Galileo as believing that truth should belong to everyone, no matter the consequences. The conflict between the scientist and those in power climaxes with the question, "What use would it be ... to have limitless time for research if any ignorant monk in the Inquisition could just put a ban on your thoughts?"[7]

Similarly, it is claimed that when Darwin published his "On the Origin of Species", the wife of a bishop[8] exclaimed, "My dear, let us hope it isn't true. But if it is, let us pray it does not become widely known".

The importance of these battles is that they began by seriously questioning what was considered to be "common knowledge", then developed from question to exploration, experimentation and basing answers on evidence — all the building blocks of modern science.

Science constantly drives and expands the frontiers of our knowledge and understanding of the world into regions of the unknown, unexplained and even unexpected. It is a constant endeavor of conquest, as dramatic and important as any other domain explored and described by humans. It is at the front of dynamic, ever-changing ideas. Even the lines between various scientific disciplines constantly move and blur, with the merger and development of classifications. Statistics and computer science became data science; energy research incorporates engineering, physics and chemistry; scientific research is advancing at breakneck speed due to its Gordian knot with technology.

In spite of the growing accessibility of science and technology, we still feel uneasy when standing at the frontier of the unknown. It ignites our imagination as well as our fears. This is where our storytelling is at its best — and at its worst. The frontier of the unknown is the point that gives birth to belief, myths, magic and threat, navigated successfully by power-seekers whom, in our quest to deal with uncertainty, we might place our trust in.

---

[7] "**Life of Galileo**", the 1943 play by Bertolt Brecht.

[8] I find it amusing that there is an ongoing debate not so much about the implications of the saying itself, but about who exactly said it: Was it the wife of the Bishop of Worcester? Or perhaps the wife of the Bishop of Birmingham? Maybe someone totally different? Was it said at all?

Science denial is rooted in our craving to develop a well-defined identity, one that gives us a sense of belonging, safety, comfort as well as meaning to our lives and what happens within them. It leads us to adhere to traditions, ideas and practices that build the communities we are or want to be part of.

Changing attitudes, beliefs, emotional needs and habits comes hard, and has little to do with knowledge and reasoning alone. Think about the parents who warn their daughter for pursuing a career in science instead of marrying and raising children, as their tradition requires, accusing her that it will be a harmful sign of "leaving the faith", claiming it will destroy the fabric of the entire community; or about the thriving industry of faith healers, marketing methods to "connect to spirits" of our long-gone beloved ones who we miss so much, offering us the "certainty" of card readings and "answers" to questions we are anxious to know about the future.

Our conflict has little to do with the logical, cognitive pathways of science in society. It has to do with forces that play in the emotional and social fields of human power, politics and economy. In the 1980s, Robert Pollack, dean and biologist at Columbia University, claimed that graduates

> *should be able to distinguish between evolution and creation, astronomy and astrology, legitimate cancer research and medical quackery. The alternative to scientific thought is fundamentalism — passively waiting for a mystical authority to tell you where the truth lies. We have ample reason, in the late 20th century, to fear the consequences of entrusting our destiny to fools, charlatans, and madmen".*[9]

Unfortunately, the issue Pollack warned against is yet to be solved. It is somewhat surprising that science denial is still so popular, as the COVID-19 pandemic exposed so clearly. When people still believe the earth is flat, we may dismiss them as ridiculous, but at least they're not harming anyone. Spreading fake news regarding COVID-19 is a completely different case. It's costing lives. In an age

---

[9] In "**Science: The Missing Link in General Education**", Columbia College Today, 1981.

of ubiquitous knowledge for all, people are still falling into pits of conspiracy theories.

Ubiquitous knowledge also means unlimited, sometimes overwhelming, exposure — To news, to fake news. To truths, lies, manipulations. To heroes-for-a-moment who inspire or disappoint.

Fake news and scandals are always more interesting, igniting our imagination and feeding our appetite for stories. We have become more doubtful and suspicious.

The fact that we can all send through the winds of social networks anything we like, unscreened, unchecked, untested, many times for no more than achieving our 15 minutes of glory, means that more than ever before we all need abilities of critical thinking, evidence-based decision making and responsibility.

Understanding science is hard to do.

It's easier to deny it.

As the influence of science and technology has become more evident in our lives, recent years have seen the emergence of a "new and improved" shape of science denial. It is a form that instead of *denying* science *uses* distorted scientific information and methodologies to make its claim, relying on "logic, experimentation and evidence-based thinking". It is no less than **science abuse**.

Science abuse relies on science to make its point. It claims that "*Your* science is proving *my* belief!" In 2014, for example, creationists demanded equal airtime on the science series "Cosmos"[10] that described evolution. Their claim was that the science program "wasn't balanced" without including their religious arguments, and that they should be able to present "the facts and evidence" that creationism is built on.

Putting aside that, in fact, scientific evidence against the religious creationist paradigm is staggering, creationism is rooted in religious systems, not in science systems. Each one is a totally different paradigm, with completely separate ideas, goals and methodologies. The very idea of placing the two paradigms on the

[10] "Cosmos — A Space-Time Odyssey" is a 2014 TV series narrated by Neil deGrasse Tyson.

same platform legitimizes the idea that they are comparable and therefore one can replace the other.

Similarly, at the beginning of the COVID-19 pandemic, news of the efficiency of specific herbs as a medicine against COVID-19 went viral on social networks. Including "personal stories that were **evidence and proof**" that they indeed work, they further recommended where to purchase them. This trend was unsurprisingly accompanied by loud voices denying the efficiency of vaccinations. Some voices went as far as claiming that vaccinations in general as well as the existence of the COVID-19 pandemic itself were all a conspiracy.

Drink these leaves, and you will be healthy.

It reminds me of a chant we were taught in elementary school:

"An apple a day keeps the sickness away".

It's a nice song for children. It's not a prescription for medicine.

Enhancing the stronghold of interest groups that usually have nothing to do with science, science denial is a pathway that dwells on fear, ignorance, emotional pain and even desperation. It takes advantage of the very core of what science is based on:

- Uncertainty is a legitimate part of our lives;
- Science isn't about "the truth", it's about what we know as "the truth" today;
- Our scientific understanding can be mistaken and may therefore change;
- It comprises man-made disciplines, and has man-made flaws; and
- Like other human activities, the scientific world struggles with diversity, hierarchies, conflicts of interests and even biases.

Medicine, nutrition, climate change, nuclear energy or the origin of humans — Addressing these issues requires the collaboration of us all. Understanding the connection of science to all aspects of life is crucial for strengthening its position as the **go-to field** for responsible and sustainable development. In an era when people are still confusing opinions and facts, science literacy is the tool to prevent society from turning away from what it has to offer due to ignorance or anxiety, favoring ideas and opinions that lack evidence, data and understanding.

Here are a few examples regarding how it can be approached.

A Reachable Star[11]: Revisiting Our Engagement with Science

When I began my job as CEO of the Wolf Foundation,[12] one of the first questions I asked myself was — What is the importance of prizes? Why do we need the Wolf Prize, the Nobel Prize, the Oscars, the Rubinstein International Piano Master Competition, the Olympic games and so on? All are prizes based on proven and recognized excellence. Surely, the laureates and winners are all people who are highly motivated and dedicated to their fields. They are high achievers in their endeavors regardless of any prize we offer. So, why do we bother to give prizes to our best and brightest, who are already accomplished professionals?

I could think of only one good answer: The prizes are meant to send a clear-cut message to everyone: These are our heroes, this is the kind of excellence we trumpet, this is what we should all aspire to be. Especially to our young. And, if this is the message, then it has clear-cut messengers who have a meaningful social task: To leave their so-called "ivory towers", break the artificial boundaries between their positions and the general public and act as approachable role models for all. Without placing this role as a beacon leading the way, prizes are no more than elite banking.

Leaving it at that felt like a hugely missed opportunity. Besides, I knew I would prove to be a very poor and insufficient banker.

---

[11] Paraphrase of "The Unreachable Star" from the song "**The Impossible Dream (The Quest)**", written by Joe Darion for the 1965 Broadway musical "Man of La Mancha", based on the 17th-century Spanish novel by Miguel de Cervantes "Don Quixote".

[12] The Wolf Foundation awards the annual **Wolf Prize**, bestowed on scientists and artists from all over the world, based on merit — and only merit. I was proud to be its CEO during the years 2011–2017.

Naturally, I found the answer in the realm of my expertise: Education.

I began to develop different events aimed at achieving one main goal: To establish meaningful interactions and shared experiences between the Wolf Prize laureates and the general public, especially high school students and teachers. Not by lectures or presentations, but by actively engaging them all, laureates and participants alike, in dialogue and hands-on activities.

Students were asked to become familiar with the research topics that the laureates dedicated their lives to and come up with questions and ideas to discuss with these world-leading figures. Interestingly, most of the questions dealt with the greater good, exploring the relationship between the research areas, our lives and the influence on the environment. Questions such as these: How does the study of proteins help us understand human history and development? How is particle physics connected to my cellphone? What do chemists and artists work on together and why? What is the connection between architectural design of buildings, food security and environmental sustainability? How do mathematics, music and semiotics influence human brains? What should we consider when creating a code of ethics for scientists, policy makers, industries and others to work by?

Researchers, graduate students, artists, scientists, educators, policy makers, government figures and high school students all came together with the laureates to participate in exploration of their research, dilemmas, successes as well as failures. Each and every time, the interactions developed to more personal areas when laureates described their daily routines, related to how they combine their work with their family lives and even chatted about their hobbies and the ways they make time for them, turning the laureates into understandable, approachable real people. It was clear that the systemic combination of these professional and personal aspects created identification, inspiration and motivation, exactly what you expect a role model to create.

Every single year of the six that I was at the Wolf Foundation, the laureates and participants alike were extremely enthusiastic to participate in these activities. And, every year, I knew it worked when one of the participants, usually a high school student, would approach me and say, "Just wait and see, one day I'll receive the Wolf Prize as well!"

Becoming a "star" in a field of interest suddenly became a reachable quest.

The ivory tower came tumbling down.

More than any other human endeavor, engaging in science enables a firm basis for multifaceted, global human interactions. I was fortunate to be part of an international gathering of science educators via zoom, from Cambodia, the Philippines, Madagascar, Bosnia-Herzegovina, Kirgizstan, Albania, Vietnam, India, Nepal, Israel, Macedonia, Thailand and Russia. We spoke about the global problem called plastic.

Before diving in to the scientific properties of plastic and its effect on the environment, each participant discussed the characteristics of their particular community, including students, teachers, parents and other stakeholders. We visited habits, attitudes, feelings and the influence of self-confidence. Our road to scientific literacy began with understanding "who's who" within each environment, an understanding that is crucial for meaningful, constructive collaboration.

"Who's who" is a pillar of every human process. In Uganda, where mining for salt is an extremely harsh and laboring task, each family struggles to mine on the property that they have inherited from their grandfathers and great-grandfathers before them, using the same techniques their ancestors did. "It is my honor to continue in the footsteps of my great grandfather," a young man called Kenneth told me.

The idea of bringing innovative scientific techniques to make mining both easier and more efficient, or even of families partnering in tasks, made him laugh. "It can't work," he said and looked at me with pity. "It's not a question of efficiency. We believe our familial ancestors' spirits are watching and blessing us. If we change our ways, we might make them mad and our mining might be harmed. We need their blessing." Only after finding ways to ensure the continuity of these blessings in the eyes of the miners could we begin introducing possibilities of scientific and technological developments.

Another point to consider is the fact that science and technology have critically changed the time, space and ways available for learning. Children, teenagers and adults can study biology, math and coding from YouTube or apps; literature, music and chemistry can be learned via iPads; the best academic studies in almost every area you can think of are provided by the best experts from the best schools, accessible online, free for all. Mechanical engineering, virology, nutrition, computational science, basic and advance physics — All these and much more can be easily accessed and researched through Google, with literally millions of sources appearing in the palms of our hands.

The impact of this situation on education is already substantial, as the following example, one of many, shows: Answering the need to study chemistry in geographically remote areas, where there was a crucial lack of teachers and resources, the Davidson Institute developed a hybrid chemistry course for high school students. The course offers a combination of online lessons throughout the school year, conducted by the best chemistry teachers and several meet-ups at the institute itself, where students arrive for a 3-day camp each time, during which they work in teams to do experiments, meet leading chemistry professors and visit working labs at the Weizmann Institute of Science; at the end, the students take the state matriculation exam to receive full accreditation for their work.

When we started, many were skeptical that the learning process will enable the students to achieve sufficient scores on the state tests. Reality proved that not only is their learning meaningful but they actually achieve higher scores than their peers who study via traditional learning processes. Students have declared that the hybrid course had a two-fold success: It allowed them to adapt more personalized learning, while at the same time developing teamwork and social interactions. The Davidson Institute continued to develop similar hybrid courses in other subject areas, such as earth sciences, physics and neurobiology, as well as many other unconventional programs to engage people of all ages in science.

There is widespread lack of agreement regarding what scientific literacy should include. Relevant questions are as follows: What should the scope be? How do we separate the wheat from the chaff? How do we lead the development of meaningful understanding without losing the magnitude of breadth, depth and true implications of our achievements? What should the balance between knowledge, methodologies and skills be? How do we ensure relevance without falling into traps of being anecdotal and trendy? What is the place of other fields that are inseparable from science, such as technology, computer science and mathematics? These are but a few of the issues that are part of a real, deep and ongoing debate.

Finding ways to adapt science literacy to an agile, boundary-breaking and swiftly changing reality is an adventure of its own. In every issue or subject that we decide to deal with, using the following roadmap of questions proves to be an inspirational and practical source of support:

- Are the interactions between science, technology and society relevant and clear to all participants?
- Does meaningful engagement with the subject involve exploring the history and philosophy of humanity in general and of science and technology in particular?
- Does the process enhance the participants' curiosity and abilities to create, design, solve problems, develop their personal learning process as well as teamwork and other social skills?
- Is the content attractive and meaningful to the participants in ways that strengthen their sense of involvement and responsibility toward their communities, human society and the environment at large?
- Can the learning skills be generalized and transferred to other issues and subject areas?

Scientific literacy is the best tool we have to understand and function logically in the world. Combined with everything we have already pondered — our technological intelligence, the big five and education — it's the greatest cause and effect of what makes us human: The most unique, dynamic and impactful species of all.

***

"Push!"

The OB-GYN, a smiling woman, who less than an hour earlier clapped her hands and chirped "Let's go deliver some babies", now looked focused and determined. Brook, pregnant with twins, was in the advanced stages of labor.

Excited, I stood on her right side, my son stood on her left, while we both tried to hold her hands, wipe her sweat and concentrate on helping in any way we could.

Brook pushed. Then, again.

A beautiful baby girl rushed into the world and started to cry with all the strength she had in her tiny lungs. She looked almost insulted when she was placed in the arms of her overjoyed father, my son.

We all started cooing at her when the OB-GYN said, "OK, you've had your fun, now come back — baby B is on the way!"

Three pushes later, another baby girl was happy to introduce her own cries next to her sister's. "Baby B" was placed in the arms of *her* father.

My son's partner.

I am probably one of the very few people on earth who can say that, for almost 9 months, I walked around with a photo that showed two embryos side by side — my two about-to-be grandchildren (Photo 112, p. 220). Calling them "Baby A" and "Baby B", all we knew was that one was the genetic daughter of my son and the other was the genetic daughter of his partner. They had the same egg donor, so the babies were half-sisters genetically on the female side of the bargain.

On the male side, however, it was a different story. The eggs were placed in two separate dishes, and each dish was fertilized with the sperm of one of the men. Several embryos developed in each dish, and the strongest from each was transferred and implanted into the womb of the surrogate.

Nine months later, two baby girls were born.

It was a very emotional, happy sight: Two fathers, two daughters — one for each biological father — one egg donor who wasn't physically present, and one surrogate.

The minute I saw the girls, I started laughing. They were no longer babies "A" and "B". Each was the spitting image of her genetic father. Natural biology, or more specifically, genetics, at its best.

But, in this case, natural biology was just part of the picture. It needed some help.

As I watched the scene in front of me, I had a few insights:

No other organism can do this.
Only a few years earlier, nothing like this could have been possible, even for humans.
Now, thanks to human ingenuity, it was a huge privilege to be present at the birth of the daughters of my son and his partner.
My science and technology granddaughters.
The most exciting, important achievement humanity can offer.
And with this, I rest my case.

Afterword

We have explored some of the main characteristics of our species. We made the shocking discovery that we are but one of many animals, an integral part of a huge and diverse array of living creatures, that, at least in my opinion, are inspiring and powerful as they are vulnerable.

We then reassured ourselves that we are indeed a very unique organism. Our uniqueness isn't a result of "*the* secret of success of humans", as is usually coined. Instead, it is a result of an entire **system** of factors, working together as a beautiful, exciting being: us.

The secrets of our uniqueness include the big five, a platform of intertwined traits that have co-evolved in our species; our sophisticated intelligence; our extremely developed technological intelligence; our inclination to educate; and our modern scientific methods and knowledge.

All this together enables us to defy our biological boundaries of time, space and physique, expanding them to extents that no other organism can achieve.

Wow!

If you have come this far, I hope you have more questions and thoughts that need further exploration than answers. I know I do. For example, we haven't explored the connection between the big five with the realms of art, music and other beautiful human practices. We haven't mentioned their connection to the dark, vulnerable, cruel and even violent side of our species. We haven't addressed the connection between our advancing technologies and the pressing issue of growing global socio-economic gaps. We haven't discussed what the next steps in education should be, on the very practical level, nor how to best connect them to the different aspects of our unique abilities.

More than anything, we haven't discussed if homo sapiens, this ultra-complex, multifaceted, inconceivable species we belong to, are indeed **a success**. There is a reason for it. While I relate to some of the positive as well as dangerous outcomes of our abilities, interpreting them as successful or not is in the eyes of the beholder.

As for me, I still enjoy toying with the quote by Charles M. Schultz with which we began our journey: "I love *mankind.* It's *people* I can't stand!"

Forever a cockeyed optimist,[1] my journey has helped me bridge the gap between the two.

The bridge you build by your journey is completely up to you.

[1] How can I end without one last footnote: **A cockeyed optimist** is a song by Oscar Hammerstein II and Richard Rodgers, from the 1949 musical "South Pacific".

Further Reading

As most readers know, it is customary to give a long, very long, sometimes *extremely* long, list of references at the end of each chapter or book that is worthy of its name. The list shows that the writer has thoroughly researched the topics discussed, and is "standing on the shoulders of others" in the arguments made.

However, in this age of ubiquitous, available and free-for-all information and knowledge, any list I compile will become more and more outdated by the minute. There are more books, articles, videos and discussions that deal with the various aspects I have written about than can be cited, and every citation will obviously lead to many others. Rest assured that, had I created such a list, it would be judged by two groups: those who find themselves in the list will say I did my research (and if they don't like my arguments, they will declare I didn't understand what they have to say), while the absentees will say I am an amateur who didn't bother to read the most important texts in the field, whatever that field may be. Furthermore, I admit that I have no desire to name-drop just to impress or show I know what I'm talking about, or in creating an endless list of references that no one will read.

So, instead of sending you to thumb through hundreds of references, let's do this: If my book has caught your interest, it will be very easy to find the relevant references it is based on, others to expand it and yet others that are blessed counter-arguments. They all encourage thoughtful discussion and debate. That is what critical thinking is all about. Any of the below resources will lead the interested reader to more information, displaying a rich fountainhead of human knowledge, fit for every flavor and every heart's content. It will also lead you to discussions displaying a variety of theories and approaches, including disagreements or support. That's exactly what they're here for: a source for further discussion and development of human thought. To achieve this goal, the following list of recommended resources can be a good start. Dive in, explore, expand — and enjoy the stories they provide. It's what makes us human.

Chapter 1: Origins

Aplin, L.M. (2019), Culture and cultural evolution in birds: A review of the evidence. *Anim Behav,* 147, ISSN 0003-3472. https://doi.org/10.1016/j.anbehav.2018.05.001.

Bekoff, M. (2010), *The Emotional Lives of Animals: A Leading Scientist Explores Animal Joy, Sorrow, and Empathy — and Why They Matter.* New World Library.

Carlson, A.D. & Copeland, J. (1985), Flash communication in fireflies. *Q Rev Biol,* 60(4).

de Waal, F. (2016), *Are We Smart Enough to Know How Smart Animals Are?* W.W. Norton & Company.

Diamond, J. (1997), *Why is Sex Fun: The Evolution of Human Sexuality.* Basic Books.

Eckert, J., Winkler, S.L., & Cartmill, E.A. (2020), Just kidding: The evolutionary roots of playful teasing. *Biol Lett,* 16, 20200370. http://dx.doi.org/10.1098/rsbl.2020.0370.

Frankl, V. (1946), *Man's Search for Meaning.* Verlag für Jugend und Volk (Austria).

González Galli, L.M. & Meinardi, E.N. (2011), The role of teleological thinking in learning the Darwinian model of evolution. *Evo Edu OutreachI,* 4, 145–152. https://doi.org/10.1007/s12052-010-0272-7.

Milo, D. (2009), *The Invention of Tomorrow.* Hkibutz Hameuhad (in Hebrew).

Olsen, L.M., Choffnes, E.R., & Mack, A. (2012), *The Social Biology of Microbial Communities: Workshop Summary.* Washington, DC: The National Academies Press.

Osvath, M. & Martin-Ordas, G. (2014), The future of future-oriented cognition in non-humans: Theory and the empirical case of the great apes. *Philos Trans R Soc Lond B Biol Sci*, 5(369), (1655), 20130486.

Strycker, N. (2014), *The Thing with Feathers*. NY: Riverhead Books.

Secor, P.R. & Dandekar, A.A. (2020), More than simple parasites: The sociobiology of bacteriophages and their bacterial hosts. *mBiol*, 11(2), e00041-20. Doi: 10.1128/mBio.00041-20.

Shivik, J.A. (2017), *Mousy Cats and Sheepish Coyotes: The Science of Animal Personalities*. Beacon Press.

Tarnita, C.E. (2017), The ecology and evolution of social behavior in microbes. *J Exp BiolI*, 220, 18–24. Doi: 10.1242/jeb.145631.

Thiery, S. & Kaimer, C. (2020), The predation strategy of *Myxococcus xanthus*. *Front Microbiol*, 11, 2. Doi: 10.3389/fmicb.2020.00002.

Van Leeuwen, E.J.C., Cronin, K.A., & Haun, D.B.M. (2014), A group-specific arbitrary tradition in chimpanzees. *Anim Cogn*, 17, 1421–1425. https://doi.org/10.1007/s10071-014-0766-8.

Velicer, G.J. & Vos, M. (2009), Sociobiology of the myxobacteria. *Annu Rev Microbiol*, 63(1), 599–623.

Whiten, A. (2021), The burgeoning reach of animal culture. *Science*, 372, 46.

Wild, S., Hoppitt, W.J.E., Allen, S.J., & Krutzen, M. (2020), Integrating genetic, environmental and social networks to reveal transmission pathways of a dolphin foraging innovation. *Curr Biol*, 30–15, P3024–3030.E4. Doi: https://doi.org/10.1016/j.cub.2020.05.069.

Wilson, E.O. (1975), *Sociobiology: The New Synthesis*. Harvard University Press.

Zahavi, A.A. (1997), *The Handicap Principle: A Missing Piece of Darwin's Puzzle*. Oxford University Press.

Zentall, T.R. (2013), Animals represent the past and the future. *J Evolution Psychol*.

Chapter 2: The Big Five

Chan, E.K.F., Timmermann, A., Baldi, B.F. *et al.* (2019), Human origins in a southern African palaeo-wetland and first migrations. *Nature*, 575, 185–189. https://doi.org/10.1038/s41586-019-1714-1.

Chappell, J., Cutting, N., Apperly, I.A., & Beck, S.R. (2013), The development of tool manufacture in humans: What helps young children make innovative tools? *PhilosTrans Roy Soci Biol Sci*.

Deacon, T. (1998), *The Symbolic Species: The Co-Evolution of Language and the Brain.* W.W. Norton & Company.

Diamond, J. (1997), *Guns, Germs, and Steel: The Fates of Human Societies.* W.W. Norton & Company.

d'Errico, F. & Backwell, L. (2005), An overview of the patterns of behavioural change in Africa and Eurasia during the Middle and Late Pleistocene. In *From Tools to Symbols: From Early Hominids to Modern Humans.* Johannesburg: Witwatersrand University Press, pp. 294–332.

Gallagher, J. (2013), James Bond is an 'impotent drunk'. *Health and Science Reporter*, BBC News.

Greenbaum, G., Getz, W.M., Rosenberg, N.A. *et al.* (2019), Disease transmission and introgression can explain the long-lasting contact zone of modern humans and Neanderthals. *Nat Commun*, 10, 5003. https://doi.org/10.1038/s41467-019-12862-7.

Hammer, M.F., Woerner, A.E., Mendez, F.L., Watkins, J.C., & Wall, J.D. (2011), Genetic evidence for archaic admixture in Africa. *Proc Natl Acad Sci.* Doi: 10.1073/pnas.1109300108.

Luis Zaman, L., Meyer, J.R., Devangam, S., Bryson, D.M., Lenski, R.E., & Ofria, C. (2014), Coevolution drives the emergence of complex traits and promotes evolvability. https://doi.org/10.1371/journal.pbio.1002023.

Noah Harari, Y. (2011), *Sapiens: A Brief History of Humankind.* Dvir Publishing (Hebrew).

Payne, J.L. & Wagner, A. (2019), The causes of evolvability and their evolution. *Nat Rev Genet*, 20, 24–38. https://doi.org/10.1038/s41576-018-0069-z.

Pigliucci, M. (2008), Is evolvability evolvable? *Nat Rev Genet*, 9, 75–82. https://doi.org/10.1038/nrg2278.

Pobiner, B., (2016),The first butchers. *Sapiens Magazine.*

Prévost, M., Groman-Yaroslavski, I., Crater Gershtein, K.M., Tejero, J.M., & Zaidner, Y, (2021), Early evidence for symbolic behavior in the Levantine Middle Paleolithic: A 120 ka old engraved aurochs bone shaft from the open-air site of Nesher Ramla, Israel. *Quater Int*, https://doi.org/10.1016/j.quaint.2021.01.002.

Russell, B. (1967), *Why I Am Not a Christian and Other Essays on Religion and Related Subjects.* Touchstone.

Sagan, C. (1997), *The Demon-Haunted World: Science as a Candle in the Dark.* Ballantine Books.

Scerri, E.M., Chikhi, L., & Thomas, M.G., (2019), Beyond multiregional and simple out-of-Africa models of human evolution. *Nat Ecol Evol*, 3(10), 1–3.

Thompson, J.N. (2010), Four central points about coevolution. *Evo Edu Outreach*, 3, 7–13.

Tylén, K., Fusaroli, R., Rojo, S., Heimann, K., Nicolas Fay, N., Johannsen, N.N., Riede, F., & Lombard, M. (2020), The evolution of early symbolic behavior in Homo sapiens. *PNAS*, 117(9), 4578–4584. https://doi.org/10.1073/pnas.1910880117.

Van de Waal, E., Borgeaud, C., & Whiten, A. (2013), Potent social learning and conformity shape a wild primate's foraging decisions. *Science*, 340(6131), 483–485. Doi: 10.1126/science.1232769.

Zidan, Dr. B. (2020), The concept and utilization of Swastika 'hooked cross' on Islamic artefacts. *J Genl Union Arab Archaeol*, 5(1). https://digitalcommons.aaru.edu.jo/jguaa/vol5/iss1/2.

Chapter 3: It's Complicated

Chamberlain, D. (2014), *Necessary Lies*. NY: St. Martin's Griffin Press.

Damasio, A.R. (1994), *Descartes' Error*. Avon Books.

Gardner, H. (1983), *Frames of Mind: The Theory of Multiple Intelligences*. Basic Books.

Gould, S.J. (1981), *The Mismeasure of Man*. W.W. Norton & Company.

Jones, N., O'Brien, M., & Ryan, T. (2018), Representation of future generations in united kingdom policy-making. *Futures*, 10.1016/j.futures.2018.01.007.

Kelly, P.J., Lewis, J.L., & Schaefer, G. (1987), *Education and Health*. Pergamon Press.

Laferton, J.A., Kube, T., Salzmann, S., Auer, C.J., & Shedden-Mora, M.C. (2017), Patients' expectations regarding medical treatment: A critical review of concepts and their assessment. *Front Psychol*, 8, 233. https://doi.org/10.3389/fpsyg.2017.00233.

Lameira, A.R. & Call, J. (2018), Time-space–displaced responses in the orangutan vocal system. *Sci Adv*, 4(11). Doi: 10.1126/sciadv.aau3401.

Ord, T. (2020), *The Precipice: Existential Risk and the Future of Humanity*, Bloomsbury Publishing.

Reed, Henry (1942), *The Naming of Parts*. Poem: https://allpoetry.com/Naming-of-Parts.

Shurkin, J. (2014), Animals that self-medicate. *PNAS*, 111(49), 17339–17341. https://doi.org/10.1073/pnas.1419966111.

Smaers, J.B. *et al.* (2021), The evolution of mammalian brain size. *Sci Adv*, 7. Doi: 10.1126/sciadv.abe2101.

Smith, T.M., Tafforeau, P., J. Reid, D.J., Grün, R., Eggins, S., Boutakiout, M., & Hublin, J.J. (2007), Earliest evidence of modern human life history in North African early Homo sapiens. *Proc Natl Acad Sci*, 104(15), 6128–6133.

Sternberg, R. (1996), *Successful Intelligence.* Simon & Schuster.

Sternberg, R. (2021), *Adaptive Intelligence.* Cambridge University Press.

Yang, C., Ye, P., Huo, J., Moller, A.P., Liang, W., & Feeney, W.E. (2020), Sparrows use a medicinal herb to defend against parasites and increase offspring condition. *Curr Biol*, 30(23). https://www.cell.com/current-biology/fulltext/S0960-9822(20)31525-6.

Chapter 4: Enter, The King: Technological Intelligence

Arthur, W.B. (2009), *The Nature of Technology: What It Is and How It Evolves.* New York: Free Press.

Beck, B.B. (1980), *Animal Tool Behavior: The Use and Manufacture of Tools.* New York: Garland STPM.

Berna, F., Goldberg, P., Horwitz, L.K., Brink, J., Holt, S., Bamford, M., & Chazan, M. (2012), *Microstratigraphic Evidence of in situ Fire in the Acheulean strata of Wonderwerk Cave.* N Wrangham R.

Brockman, J. (1995), *The Third Culture: Beyond the Scientific Revolution.* Simon & Schuster.

Carmody, R.N., Dannemann, M., Briggs, A.W., Nickel, B., Groopman, E.E., Wrangham, R.W., & Kelso, J. (2016), Genetic evidence of human adaptation to a cooked diet. *Genome Biology and Evolution.* https://doi.org/10.1093/gbe/evw059.

Ellul, J., (1964), *The Technological Society,* Vintage Books. Albert A. Knopf, Inc. & Random House, Inc.

Ellul, J. (1980), *The Technological System — 2018 Edition.* Wipf & Stock Publishers.

Geva-Sagiv, M., Las, L., Yovel, Y., & Ulanovsky, N. (2015), Spatial cognition in bats and rats: From sensory acquisition to multiscale maps and navigation. *Nat Rev Neurosci*, 16, 94–108. https://doi.org/10.1038/nrn3888.

Gowlett, J.A.J. & Wrangham, R.W. (2013), Earliest fire in Africa: Towards the convergence of archaeological evidence and the cooking hypothesis. *Azania: Archaeol Res Africa,* 48(1), 5–30. Doi: 10.1080/0067270X.2012.756754.

Hiscock, H.G. *et al.* (2016), The quantum needle of the avian magnetic compass. *PNAS,* 113(17). https://doi.org/10.1073/pnas.1600341113.

Hopkin, M. (2007), Chimps make spears to catch dinner. *Nature,* https://doi.org/10.1038/news070219-11.

Johnson, B. (xxxx), *The Great Horse Manure Crisis of 1894, Historic UK.* https://www.historic-uk.com/HistoryUK/HistoryofBritain/Great-Horse-Manure-Crisis-of-1894/.

Kurzweil, R. (2006), *The Singularity Is Near: When Humans Transcend Biology.* Viking.

Mann, J. *et al.* (2008), Why do dolphins carry sponges? *PloS One,* 3(12), e3868. Doi:10.1371/journal.pone.0003868.

Minsky, M. (1988), *The Society of Mind,* Touchstone Edition. Simon & Schuster, Inc.

Ruhl, C. (2020), *Intelligence: Definition, Theories and Testing. Simply Psychology.* https://www.simplypsychology.org/intelligence.html.

Schatzberg, E. (2018), *Technology: Critical History of a Concept.* Chicago and London: University of Chicago Press.

Snow, C.P. (1959), *The Two Cultures.* Cambridge University Press.

Sobel, D. (1995), *Longitude.* New York: Walker & Company.

St Amant, R.E. & Horton, T. (2008), Revisiting the definition of animal tool use. *Anim Behav,* 75(4).

Tenger-Trolander, A., Lu, W., Noyes, M., & Kronforst, M.R. (2019), Contemporary loss of migration in monarch butterflies. *PNAS,* 16116(29), 14671–14676.

Tsoar, A., Nathan, R., Bartan, Y., Vyssotski, A., Dell'Omo, G., & Ulanovsky, N. (2011), Large-scale navigational map in a mammal. *PNAS,* 108(37). https://doi.org/10.1073/pnas.1107365108.

Turkle, S. (2011) *Alone Together.* Basic Books.

Von Bertalanffy, L. (1968), *General System Theory: Foundations, Development, Applications.* New York: George Braziller.

Whitehead, H., Smith, T.D., & Rendell, L. (2021), Adaptation of sperm whales to open-boat whalers: Rapid social learning on a large scale? *Biol Lett,* 17, 20210030.

Wrangham, R. (2009), *Catching Fire: How Cooking Made Us Human.* New York: Basic Books.

Chapter 5: The Fifth Freedom

Ackoff, R. & Greenberg, D. (2008), *Turning Learning Right Side Up*. Pearson Education Inc. Wharton School Publishing.

Brand, S. (1999), *The Clock of the Long Now*. Basic Books.

Brunner, J. (1996), *The Culture of Education*. Harvard University Press.

Egan, K. (2008), *The Future of Education*. Yale University Press.

Leadbeater, E., Raine, N.E., & Chittka, L. (2006), Social learning: Ants and the meaning of teaching. *Curr Biol*, 16(9).

Jowett, B. (1905), *Aristotle's Politics, Book VIII*. Oxford: The Clarendon Press.

Klump, B.C. *et al.* (2021), Innovation and geographic spread of a complex foraging culture in an urban parrot. *Science*, 373(6553). Doi: 10.1126/science.abe7808.

Robinson, K. & Aronica, L. (2015), *Creative Schools: The Grassroots Revolution That's Transforming Schools*. New York: Viking.

Sizer, N.F. & Sizer, T. (2000), *The Students Are Watching*. Beacon Press.

Tishman, S., Perkins, D., & Jay, E. (1994), *The Thinking Classroom: Learning and Teaching in a Culture of Thinking*. Harvard University press.

Thomas, D. & Brown, J.S. (2011), *A New Culture of Learning*. Thomas and Brown.

Tyack, D. & Cuban, L. (1995), *Tinkering Toward Utopia: A Century of Public School Reform*. Harvard University Press.

Chapter 6: Everything is Science

Hawking, S. (2018), *Brief Answers to the Big Questions*. John Murray Publishers.

Kuhn, T.S. (2012), *The Structure of Scientific Revolutions — 50th Anniversary Edition*. University of Chicago Press.

Livio, M. (2020), *Galileo and the Science Deniers*. Simon & Schuster.

Oreskes, N. & Conway, E.M. (2010), *Merchants of Doubt*. Bloomsbury Press.

Pollack, R. (1981), *Science: The Missing Link in General Education*. Columbia College Today.

Potvin, P. & Hasni, A. (2014), Analysis of the Decline in interest towards school science and technology from grades 5 through 11. *J Sci Educ Technol* 23, 784–802. https://doi.org/10.1007/s10956-014-9512-x.

Sithole *et al.* (2017), Student attraction, persistence and retention in STEM programs: Successes and continuing challenges. *Higher Education Studies*. 7(1).

UNESCO: Functional Literacy (1978), https://learningportal.iiep.unesco.org/en/glossary/functional-literacy.

Williams, W.F. (1991), Teaching science and technology to non-science majors — The STS approach. *Bull Sci Tech Soc*, (11).

Wulf, A. (2015), *The Invention of Nature*. Alfred A. Knopf Publishing Company.

General Reading — Recommendations

Dawkins, R. (2009), *The Greatest Show on Earth*. Free Press.

Diamond, J. (1997), *Guns, Germs and Steel: The Fates of Human Societies*. Norton.

Suzanne, S. (2021), *Finding the Mother Tree*. Knopf Publisher.

Rutherford, A. (2017), *A Brief History of Everyone Who Ever Lived*. UK: Orion Publishing.

Sapolsky, R.M. (2018), *Behave: The Biology of Humans at Our Best and Worst*. Penguin Books.

Yong, E. (2016), *I Contain Multitudes*. Ecco HarperCollins Publishers.

You are invited to watch my TEDx Talks:

1. The preeminence of man/Dr. Liat Ben David, 2016. https://www.youtube.com/watch?v=vqlDdh_F8cw&t=3s&ab_channel=TEDxTalks.
2. Rerouting education — the next human revolution is here/Dr. Liat Ben David, 2017. https://www.youtube.com/watch?v=gI31_VYk2vs&ab_channel=TEDxTalks.

Photos

Chapter 1

1: Antarctica, the white continent of the South Pole

2: Antarctica, the white continent of the South Pole

3: Antarctica, the white continent of the South Pole

4: Antarctica, the white continent of the South Pole

5: Albatross courtship dance, Falkland Islands

6: Albatross courtship dance, Falkland Islands

7: Albatross courtship dance, Falkland Islands

8: Nazca Boobies courtship clappering, Galapagos Islands

9: Frigates males displaying inflated red throat pouches, Galapagos Islands

10: Frigates males displaying inflated red throat pouches, Galapagos Islands

11: Bachelor Giraffes necking, Botswana

12: Bachelor Giraffes necking, Botswana

13: Bachelor Giraffes necking, Botswana

14: Bachelor Giraffes necking, Botswana

15: Birth of Sea Lion Cub, threatened by Skuas, South Georgia Islands

16: Birth of Sea Lion Cub, threatened by Skuas, South Georgia Islands

17: Birth of Sea Lion Cub, threatened by Skuas, South Georgia Islands

18: Birth of Sea Lion Cub, threatened by Skuas, South Georgia Islands

19: Birth of Sea Lion Cub, threatened by Skuas, South Georgia Islands

20: Nests of African Weavers, Uganda

21: Nests of African Weavers, Uganda

22: Nests of African Weavers, Uganda

23: Sociable Weaver's nest, Botswana

24: Dolphins playing, Galapagos Islands

25: Chin Strapped Penguins building nests — and stealing, Antarctica

26: Chin Strapped Penguins building nests — and stealing, Antarctica

27: Elephants mourning and elephant-killing lions, Savuti, Botswana

28: Elephants mourning and elephant-killing lions, Savuti, Botswana

29: A couple of Bee-Eaters, one with an insect in his beak, Kenya

30: "Baby" King Penguin complaining to Mom, South Georgia Islands

31: Orange Diademed Sifaka lemurs playing, Madagascar

32: Orange Diademed Sifaka lemurs playing, Madagascar

33: Orange Diademed Sifaka lemurs playing, Madagascar

34: Lemurs playing, Indri enjoying a handout, Madagascar

35: Lemurs playing, Indri enjoying a handout, Madagascar

36: Lemurs playing, Indri enjoying a handout, Madagascar

37: "Mr. Lonely" King Penguin, Falkland Islands

38: "Mr. Lonely" King Penguin, Falkland Islands

39: "Mr. Lonely" King Penguin, Falkland Islands

40: Gentoo Penguins “skiing”, Antarctica

41: Gentoo Penguins “skiing”, Antarctica

42: Baboons enjoying sliding on a solar panel, Uganda

Chapter 2

44: Worship Spiritual Tree, Madagascar

45: "Lungta" flags and ritual, Tibet, China

46: "Lungta" flags and ritual, Tibet, China

47: "Lungta" flags and ritual, Tibet, China

48: Fatima, Portugal

49: Swastika symbolizing peace, Taiwan

50: Swastika symbolizing peace, Taiwan

51: Nazi memorabilia, Denmark

52: Nazi memorabilia, Denmark

53: Endless riches of the Galapagos Islands

54: Endless riches of the Galapagos Islands

55: Endless riches of the Galapagos Islands

56: Endless riches of the Galapagos Islands

57: Endless riches of the Galapagos Islands

58: Endless riches of the Galapagos Islands

59: Endless riches of the Galapagos Islands

60: Endless riches of the Galapagos Islands

61: Endless riches of the Galapagos Islands

Chapter 3

62: Curious seals, South Georgia Islands

63: Curious seals, South Georgia Islands

64: Curious penguins, South Georgia Islands

65: Seals playing, South Georgia Islands

66: Standing in the way of penguins, South Georgia Islands

67: Standing in the way of penguins, South Georgia Islands

68: Standing in the way of penguins, South Georgia Islands

69: Standing in the way of penguins, South Georgia Islands

70: Standing in the way of penguins, South Georgia Islands

71: Standing in the way of penguins, South Georgia Islands

72: Standing in the way of penguins, South Georgia Islands

73: What is he thinking about? A Chimp deep in thought, Uganda

74: Big body, small head, intelligent seal, Galapagos Islands

75: Agriculture by deliberate fire, Botswana

76: Agriculture by deliberate fire, Botswana

77: Elephants at the edge of a burnt field, Botswana

78: Elephants at the edge of a burnt field, Botswana

79: Elephants at the edge of a burnt field, Botswana

Chapter 4

80: Dancing Sifaka Lemurs, Madagascar

81: Dancing Sifaka Lemurs, Madagascar

82: Dancing Sifaka Lemurs, Madagascar

83: Dancing Sifaka Lemurs, Madagascar

84: Dancing Sifaka Lemurs, Madagascar

85: Dancing Sifaka Lemurs, Madagascar

86: Dancing Sifaka Lemurs, Madagascar

87: Temples of Tikal, Guatemala

88: Temples of Tikal, Guatemala

89: Temples of Tikal, Guatemala

90: Solar panels on mud huts, Ethiopia and Uganda

91: Solar panels on mud huts, Ethiopia and Uganda

92: Grytviken Bay, South Georgia Islands

93: Grytviken Bay, South Georgia Islands

94: Grytviken Bay, South Georgia Islands

95: Grytviken Bay, South Georgia Islands

96: Grytviken Bay, South Georgia Islands

97: Grytviken Bay, South Georgia Islands

98: Grytviken Bay, South Georgia Islands

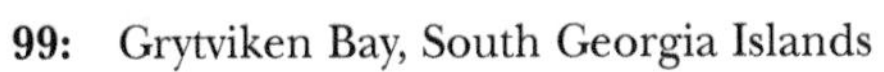

99: Grytviken Bay, South Georgia Islands

Chapter 5

100: Schools are similar and easily recognized throughout the world, here in Cuba.

101: School in Thailand

102: School in Norway (Svalbard)

103: School in Madagascar

104: School in Madagascar

105: School in Costa Rica

106: School in Madagascar

107: School in Uganda

Chapter 6

108: Pamela's school, Uganda

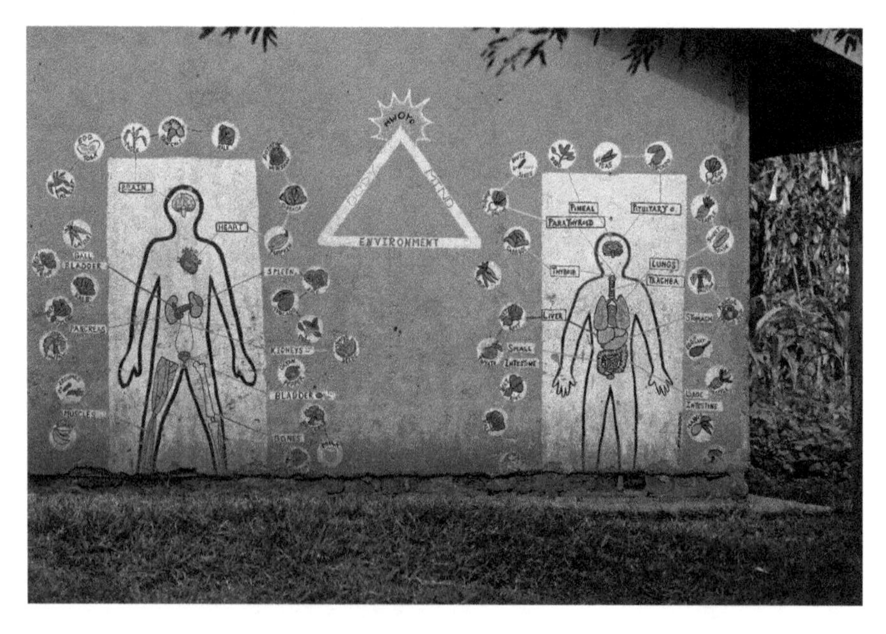

109: Pamela's school, Uganda

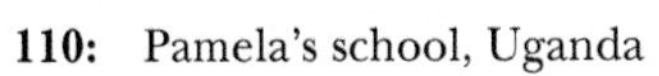

110: Pamela's school, Uganda

111: Pamela's school, Uganda

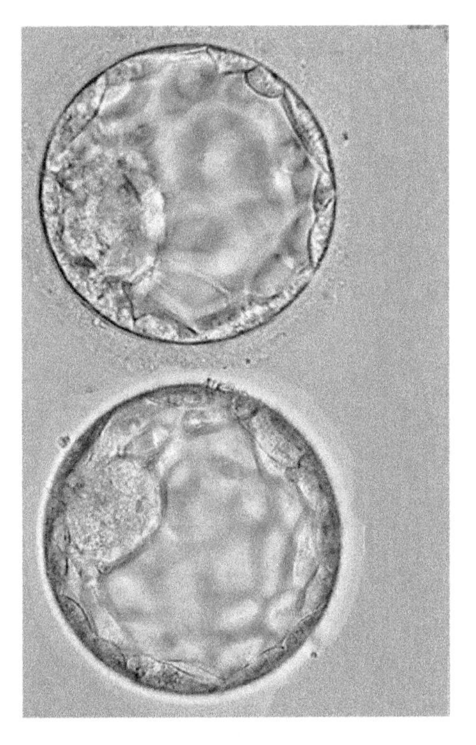

112: Two embryos, my granddaughters-to-be

Index

CPSIA information can be obtained
at www.ICGtesting.com
Printed in the USA
BVHW090857010522
635441BV00001B/3

9 789811 248528